Leveled Texts
for Mathematics
Geometry

Author

Lori Barker

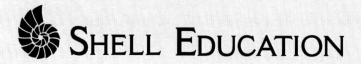

SHELL EDUCATION

Consultant

Barbara Talley, M.S.
Texas A&M University

Publishing Credits

Dona Herweck Rice, *Editor-in-Chief*; Lee Aucoin, *Creative Director*; Don Tran, *Print Production Manager*;
Sara Johnson, M.S.Ed., *Senior Editor*; Hillary Wolfe, *Editor*; Lisa Greathouse, *Editor*;
Evelyn Garcia, *Associate Education Editor*; Neri Garcia, *Cover Designer*; Juan Chavolla, *Production Artist*;
Stephanie Reid, *Photo Editor*; Corinne Burton, M.S.Ed., *Publisher*

All images from Shutterstock.com

Shell Education

5301 Oceanus Drive

Huntington Beach, CA 92649

http://www.shelleducation.com

ISBN 978-1-4258-0717-7

©2011 Shell Educational Publishing, Inc.

The classroom teacher may reproduce copies of materials in this book for classroom use only. The reproduction of any part for an entire school or school system is strictly prohibited. No part of this publication may be transmitted, stored, or recorded in any form without written permission from the publisher.

Table of Contents

What Is Differentiation? 4
How to Differentiate Using This Product 5
General Information About Student Populations 6
    Below-Grade-Level Students 6
    English Language Learners 6
    On-Grade-Level Students 7
    Above-Grade-Level Students 7
Strategies for Using the Leveled Texts 8
    Below-Grade-Level Students 8
    English Language Learners 11
    Above-Grade-Level Students 14
How to Use This Product 16
    Readability Chart 16
    Components of the Product 16
    Tips for Managing the Product 18
    Correlations to Mathematics Standards 19
Leveled Texts 21
    Angles All Around 21
    Understanding Triangles 29
    To Cross or Not to Cross 37
    Quadrilaterals 45
    Classifying 2-D Shapes 53
    Irregular Shapes 61
    Congruent and Similar Figures 69
    Understanding 3-D Shapes 77
    Understanding Prisms 85
    The Coordinate Plane 93
    Circles 101
    Symmetry 109
    Reflections 117
    Rotations 125
    Translations 133
Appendices 141
    References Cited 141
    Contents of Teacher Resource CD 142

What Is Differentiation?

Over the past few years, classrooms have evolved into diverse pools of learners. Gifted students, English language learners, special needs students, high achievers, underachievers, and average students all come together to learn from one teacher. The teacher is expected to meet their diverse needs in one classroom. It brings back memories of the one-room schoolhouse during early American history. Not too long ago, lessons were designed to be one size fits all. It was thought that students in the same grade level learned in similar ways. Today, we know that viewpoint to be faulty. Students have differing learning styles, come from different cultures, experience a variety of emotions, and have varied interests. For each subject, they also differ in academic readiness. At times, the challenges teachers face can be overwhelming, as they struggle to figure out how to create learning environments that address the differences they find in their students.

What is differentiation? Carol Ann Tomlinson at the University of Virginia says, "Differentiation is simply a teacher attending to the learning needs of a particular student or small group of students, rather than teaching a class as though all individuals in it were basically alike" (2000). Differentiation can be carried out by any teacher who keeps the learners at the forefront of his or her instruction. The effective teacher asks, "What am I going to do to shape instruction to meet the needs of all my learners?" One method or methodology will not reach all students.

Differentiation encompasses what is taught, how it is taught, and the products students create to show what they have learned. When differentiating curriculum, teachers become the organizers of learning opportunities within the classroom environment. These categories are often referred to as content, process, and product.

- **Content:** Differentiating the content means to put more depth into the curriculum through organizing the curriculum concepts and structure of knowledge.
- **Process:** Differentiating the process means using varied instructional techniques and materials to enhance students' learning.
- **Product:** When products are differentiated, cognitive development and the students' abilities to express themselves improve, as they are given different product options.

Teachers should differentiate content, process, and product according to students' characteristics, including students' readiness, learning styles, and interests.

- **Readiness:** If a learning experience aligns closely with students' previous skills and understanding of a topic, they will learn better.
- **Learning styles:** Teachers should create assignments that allow students to complete work according to their personal preferences and styles.
- **Interests:** If a topic sparks excitement in the learners, then students will become involved in learning and better remember what is taught.

How to Differentiate Using This Product

According to the Common Core State Standards (2010), all students need to learn to read and discuss concepts across the content areas in order to be prepared for college and beyond. The leveled texts in this series help teachers differentiate mathematics content for their students to allow all students access to the concepts being explored. Each book has 15 topics, and each topic has a text written at four different reading levels. (See page 17 for more information.) While these texts are written at a variety of reading levels, all the levels remain strong in presenting the mathematics content and vocabulary. Teachers can focus on the same content standard or objective for the whole class, but individual students can access the content at their instructional reading levels rather than at their frustration levels.

Determining your students' instructional reading levels is the first step in the process. It is important to assess their reading abilities often so they do not get tracked into one level. Below are suggested ways to determine students' reading levels.

- **Running records:** While your class is doing independent work, pull your below-grade-level students aside, one at a time. Have them read aloud the lowest level of a text (the star level) individually as you record any errors they make on your own copy of the text. If students read accurately and fluently and comprehend the material, move them up to the next level and repeat the process. Following the reading, ask comprehension questions to assess their understanding of the material. Use your judgment to determine whether students seem frustrated as they read. As a general guideline, students reading below 90% accuracy are likely to feel frustrated as they read. There are also a variety of published reading assessment tools that can be used to assess students' reading levels with the running record format.

- **Refer to other resources:** Other ways to determine instructional reading levels include checking your students' Individualized Education Plans (IEP), asking the school's resource teachers, or reviewing test scores. All of these resources should be able to give you the additional information you need to determine the reading level to begin with for your students.

Teachers can also use the texts in this series to scaffold the content for their students. At the beginning of the year, students at the lowest reading levels may need focused teacher guidance. As the year progresses, teachers can begin giving students multiple levels of the same text to allow them to work independently to improve their comprehension. This means each student would have a copy of the text at his or her independent reading level and instructional reading level. As students read the instructional-level texts, they can use the lower texts to better understand the difficult vocabulary. By scaffolding the content in this way, teachers can support students as they move up through the reading levels. This will encourage students to work with texts that are closer to the grade level at which they will be tested.

General Information About the Student Populations

Below-Grade-Level Students

By Dennis Benjamin

Gone are the days of a separate special education curriculum. Federal government regulations require that special needs students have access to the general education curriculum. For the vast majority of special-needs students today, their Individualized Education Plans (IEPs) contain current and targeted performance levels but few short-term content objectives. In other words, the special-needs students are required to learn the same content as their on-grade-level peers.

Be well aware of the accommodations and modifications written in students' IEPs. Use them in your teaching and assessment so they become routine. If you hold high expectations of success for all of your students, their efforts and performances will rise as well. Remember the root word of *disability* is *ability*. Go to the root needs of the learner and apply good teaching. The results will astound and please both of you.

English Language Learners

By Marcela von Vacano

Many school districts have chosen the inclusion model to integrate English language learners into mainstream classrooms. This model has its benefits as well as its drawbacks. One benefit is that English language learners may be able to learn from their peers by hearing and using English more frequently. One drawback is that these second-language learners cannot understand academic language and concepts without special instruction. They need sheltered instruction to take the first steps toward mastering English. In an inclusion classroom, the teacher may not have the time or necessary training to provide specialized instruction for these learners.

Acquiring a second language is a lengthy process that integrates listening, speaking, reading, and writing. Students who are newcomers to the English language are not able to process information until they have mastered a certain number of structures and vocabulary words. Students may learn social language in one or two years. However, academic language takes up to eight years for most students.

Teaching academic language requires good planning and effective implementation. Pacing, or the rate at which information is presented, is another important component in this process. English language learners need to hear the same word in context several times, and they need to practice structures to internalize the words. Reviewing and summarizing what was taught are absolutely necessary for English language learners.

General Information About the Student Populations (cont.)

On-Grade-Level Students

By Wendy Conklin

On-grade-level students often get overlooked when planning curriculum. More emphasis is usually placed on those who struggle and, at times, on those who excel. Teachers spend time teaching basic skills and even go below grade level to ensure that all students are up to speed. While this is a noble endeavor and is necessary at times, in the midst of it all, the on-grade-level students can get lost in the shuffle. We must not forget that differentiated strategies are good for the on-grade-level students, too. Providing activities that are too challenging can frustrate these students; on the other hand, assignments that are too easy can be boring and a waste of their time. The key to reaching this population successfully is to find just the right level of activities and questions while keeping a keen eye on their diverse learning styles.

Above-Grade-Level Students

By Wendy Conklin

In recent years, many state and school district budgets have cut funding that has in the past provided resources for their gifted and talented programs. The push and focus of schools nationwide is proficiency. It is important that students have the basic skills to read fluently, solve problems, and grasp mathematical concepts. As a result, funding has been redistributed in hopes of improving test scores on state and national standardized tests. In many cases, the attention has focused only on improving low test scores to the detriment of the gifted students who need to be challenged.

Differentiating the products you require from your students is a very effective and fairly easy way to meet the needs of gifted students. Actually, this simple change to your assignments will benefit all levels of students in your classroom. While some students are strong verbally, others express themselves better through nonlinguistic representation. After reading the texts in this book, students can express their comprehension through different means, such as drawings, plays, songs, skits, or videos. It is important to identify and address different learning styles. By giving more open-ended assignments, you allow for more creativity and diversity in your classroom. These differentiated products can easily be aligned with content standards. To assess these standards, use differentiated rubrics.

Strategies for Using the Leveled Texts

Below-Grade-Level Students

By Dennis Benjamin

Vocabulary Scavenger Hunt

A valuable prereading strategy is a Vocabulary Scavenger Hunt. Students preview the text and highlight unknown words. Students then write the words on specially divided pages. The pages are divided into quarters with the following headings: *Definition*, *Sentence*, *Examples*, and *Nonexamples*. A section called *Picture* can be placed over the middle of the chart to give students a visual reminder of the word and its definition.

Example Vocabulary Scavenger Hunt

estimate

Definition	Sentence
to determine roughly the size or quantity	We need to estimate how much paint we will need to buy.
Examples	**Nonexamples**
estimating a distance; estimating an amount	measuring to an exact number; weighing and recording an exact weight

This encounter with new vocabulary enables students to use it properly. The definition identifies the word's meaning in student-friendly language. The sentence should be written so that the word is used in context. This helps the student make connections with background knowledge. Illustrating the sentence gives a visual clue. Examples help students prepare for factual questions from the teacher or on standardized assessments. Nonexamples help students prepare for ***not*** and ***except for*** test questions such as "All of these are polygons *except for*…" and "Which of these terms in this expression is not a constant?" Any information the student was unable to record before reading can be added after reading the text.

Strategies for Using the Leveled Texts (cont.)

Below-Grade-Level Students (cont.)

Graphic Organizers to Find Similarities and Differences

Setting a purpose for reading content focuses the learner. One purpose for reading can be to identify similarities and differences. This is a skill that must be directly taught, modeled, and applied. The authors of *Classroom Instruction That Works* state that identifying similarities and differences "might be considered the core of all learning" (Marzano, Pickering, and Pollock 2001). Higher-level tasks include comparing and classifying information and using metaphors and analogies. One way to scaffold these skills is through the use of graphic organizers, which help students focus on the essential information and organize their thoughts.

Example Classifying Graphic Organizer

Equation	Constants	Variables	Number of Terms
$7 + 12 = 19$	7, 12, 19	none	3
$3x = 12$	12	x	2
$a + b$	none	a, b	2
$a^2 + b^2 + c^2$	none	a, b, c	3

The Riddles Graphic Organizer allows students to compare and contrast two-dimensional shapes using riddles. Students first complete a chart you've designed. Then, using that chart, they can write summary sentences. They do this by using the riddle clues and reading across the chart. Students can also read down the chart and write summary sentences. With the chart below, students could write the following sentences: A circle is not a polygon. The interior angles of a triangle always add up to 180°.

Example Riddles Graphic Organizer

What Am I?	Circle	Square	Triangle	Rectangle
I come in different configurations.			x	x
I am a polygon.		x	x	x
I am a closed shape.	x	x	x	x
My interior angles always add up to 180°.			x	
I have at least three vertices.		x	x	x

Strategies for Using the Leveled Texts (cont.)

Below-Grade-Level Students (cont.)

Framed Outline

This is an underused technique that bears great results. Many below-grade-level students have problems with reading comprehension. They need a framework to help them attack the text and gain confidence in comprehending the material. Once students gain confidence and learn how to locate factual information, the teacher can fade out this technique.

There are two steps to successfully using this technique. First, the teacher writes cloze sentences. Second, the students complete the cloze activity and write summary sentences.

Example Framed Outline

A _____ graph is used to show how two variables may be related to each other. A graph should have a _____. The axes should be labeled and a proper scale should be shown.

Summary Sentences

A good graph should correctly show the data. It should have a title; the axes should be labeled correctly; it should show the proper scale.

Modeling Written Responses

A frequent criticism heard by special educators is that below-grade-level students write poor responses to content-area questions. This problem can be remedied if resource and classroom teachers model what good answers look like. While this may seem like common sense, few teachers take the time to do this. They just assume all children know how to respond in writing.

This is a technique you may want to use before asking your students to respond to the You Try It questions associated with the leveled texts in this series. First, read the question aloud. Then, write the question on the board or an overhead and think aloud about how you would go about answering the question. Next, solve the problem showing all the steps. Introduce the other problems and repeat the procedure. Have students explain how they solved the problems in writing so that they make the connection that quality written responses are part of your expectations.

Strategies for Using the Leveled Texts (cont.)

English Language Learners

By Marcela von Vacano

Effective teaching for English language learners requires effective planning. In order to achieve success, teachers need to understand and use a conceptual framework to help them plan lessons and units. There are six major components to any framework. Each is described in detail below.

1. Select and Define Concepts and Language Objectives—Before having students read one of the texts in this book, the teacher must first choose a mathematical concept and language objective (reading, writing, listening, or speaking) appropriate for the grade level. Then, the next step is to clearly define the concept to be taught. This requires knowledge of the subject matter, alignment with local and state objectives, and careful formulation of a statement that defines the concept. This concept represents the overarching idea. The mathematical concept should be written on a piece of paper and posted in a visible place in the classroom.

By the definition of the concept, post a set of key language objectives. Based on the content and language objectives, select essential vocabulary from the text. The number of new words selected should be based on students' English language levels. Post these words on a word wall that may be arranged alphabetically or by themes.

2. Build Background Knowledge—Some English language learners may have a lot of knowledge in their native language, while others may have little or no knowledge. The teacher will want to build the background knowledge of the students using different strategies such as the following:

> **Visuals:** Use posters, photographs, postcards, newspapers, magazines, drawings, and video clips of the topic you are presenting.
>
> **Realia:** Bring real-life objects to the classroom. If you are teaching about measurement, bring in items such as thermometers, scales, time pieces, and rulers.
>
> **Vocabulary and Word Wall:** Introduce key vocabulary in context. Create families of words. Have students draw pictures that illustrate the words and write sentences about the words. Also, be sure you have posted the words on a word wall in your classroom.
>
> **Desk Dictionaries:** Have students create their own desk dictionaries using index cards. On one side, they should draw a picture of the word. On the opposite side, they should write the word in their own language and in English.

Strategies for Using the Leveled Texts (cont.)

English Language Learners (cont.)

3. Teach Concepts and Language Objectives—The teacher must present content and language objectives clearly. He or she must engage students using a hook and must pace the delivery of instruction, taking into consideration students' English language levels. The concept or concepts to be taught must be stated clearly. Use the first languages of the students whenever possible or assign other students who speak the same languages to mentor and to work cooperatively with the English language learners.

Lev Semenovich Vygotsky, a Russian psychologist, wrote about the Zone of Proximal Development (ZPD). This theory states that good instruction must fill the gap that exists between the present knowledge of a child and the child's potential (1978). Scaffolding instruction is an important component when planning and teaching lessons. English language learners cannot jump stages of language and content development. You must determine where the students are in the learning process and teach to the next level using several small steps to get to the desired outcome. With the leveled texts in this series and periodic assessment of students' language levels, teachers can support students as they climb the academic ladder.

4. Practice Concepts and Language Objectives—English language learners need to practice what they learn with engaging activities. Most people retain knowledge best after applying what they learn to their own lives. This is definitely true for English language learners. Students can apply content and language knowledge by creating projects, stories, skits, poems, or artifacts that show what they learned. Some activities should be geared to the right side of the brain. For students who are left-brain dominant, activities such as defining words and concepts, using graphic organizers, and explaining procedures should be developed. The following teaching strategies are effective in helping students practice both language and content:

> **Simulations:** Students learn by doing. For example, when teaching about data analysis, have students do a survey about their classmates' favorite sports. First, students make a list of questions and collect the necessary data. Then, they tally the responses and determine the best way to represent the data. Lastly, students create a graph that shows their results and display it in the classroom.

> **Literature response:** Read a text from this book. Have students choose two concepts described or introduced in the text. Ask students to create a conversation two people might have to debate which concept is useful. Or, have students write journal entries about real-life ways they use these mathematical concepts.

Strategies for Using the Leveled Texts (cont.)

English Language Learners (cont.)

4. **Practice Concepts and Language Objectives** (cont.)

    Have a short debate: Make a controversial statement such as "It isn't necessary to learn addition." After reading a text in this book, have students think about the question and take a position. As students present their ideas, one student can act as a moderator.

    Interview: Students may interview a member of the family or a neighbor in order to obtain information regarding a topic from the texts in this book. For example: What are some ways you use geometry in your work?

5. **Evaluation and Alternative Assessments**—We know that evaluation is used to inform instruction. Students must have the opportunity to show their understanding of concepts in different ways and not only through standard assessments. Use both formative and summative assessments to ensure that you are effectively meeting your content and language objectives. Formative assessment is used to plan effective lessons for a particular group of students. Summative assessment is used to find out how much the students have learned. Other authentic assessments that show day-to-day progress are: text retelling, teacher rating scales, student self-evaluations, cloze testing, holistic scoring of writing samples, performance assessments, and portfolios. Periodically assessing student learning will help you ensure that students continue to receive the correct levels of texts.

6. **Home-School Connection**—The home-school connection is an important component in the learning process for English language learners. Parents are the first teachers, and they establish expectations for their children. These expectations help shape the behavior of their children. By asking parents to be active participants in the education of their children, students get a double dose of support and encouragement. As a result, families become partners in the education of their children and chances for success in your classroom increase.

You can send home copies of the texts in this series for parents to read with their children. You can even send multiple levels to meet the needs of your second-language parents as well as your students. In this way, you are sharing your mathematics content standards with your whole second-language community.

Strategies for Using the Leveled Texts (cont.)

Above-Grade-Level Students

By Wendy Conklin

Open-Ended Questions and Activities

Teachers need to be aware of activities that provide a ceiling that is too low for gifted students. When given activities like this, gifted students become bored. We know these students can do more, but how much more? Offering open-ended questions and activities will give high-ability students the opportunities to perform at or above their ability levels. For example, ask students to evaluate mathematical topics described in the texts, with questions such as "Do you think students should be allowed to use calculators in math?" or "What do you think you would need to build a two-story dog house?" These questions require students to form opinions, think deeply about the issues, and form several different responses in their minds. To questions like these, there really is no single correct answer.

The generic, open-ended question stems listed below can be adapted to any topic. There is one You Try It question for each topic in this book. Use questions or statements like the ones shown here to develop further discussion for the leveled texts.

- In what ways did…
- How might you have done this differently…
- What if…
- What are some possible explanations for…
- How does this affect…
- Explain several reasons why…
- What problems does this create…
- Describe the ways…
- What is the best…
- What is the worst…
- What is the likelihood…
- Predict the outcome…
- Form a hypothesis…
- What are three ways to classify…
- Support your reason…
- Make a plan for…
- Propose a solution…
- What is an alternative to…

Strategies for Using the Leveled Texts (cont.)

Above-Grade-Level Students (cont.)

Student-Directed Learning

Because they are academically advanced, above-grade-level students are often the leaders in classrooms. They are more self-sufficient learners, too. As a result, there are some student-directed strategies that teachers can employ successfully with these students. Remember to use the texts in this book as jump starts so that students will be interested in finding out more about the mathematical concepts presented. Above-grade-level students may enjoy any of the following activities:

- Writing their own questions, exchanging their questions with others, and grading the responses.
- Reviewing the lesson and teaching the topic to another group of students.
- Reading other nonfiction texts about these mathematical concepts to further expand their knowledge.
- Writing the quizzes and tests to go along with the texts.
- Creating illustrated timelines to be displayed as visuals for the entire class.
- Putting together multimedia presentations about the mathematical concepts.

Tiered Assignments

Teachers can differentiate lessons by using tiered assignments, or scaffolded lessons. Tiered assignments are parallel tasks designed to have varied levels of depth, complexity, and abstractness. All students work toward one goal, concept, or outcome, but the lesson is tiered to allow for different levels of readiness and performance. As students work, they build on their prior knowledge and understanding. Students are motivated to be successful according to their own readiness and learning preferences.

Guidelines for writing tiered lessons include the following:

1. Pick the skill, concept, or generalization that needs to be learned.
2. Think of an on-grade-level activity that teaches this skill, concept, or generalization.
3. Assess the students using classroom discussions, quizzes, tests, or journal entries and place them in groups.
4. Take another look at the activity from Step 2. Modify this activity to meet the needs of the below-grade-level and above-grade-level learners in the class. Add complexity and depth for the above-grade-level students. Add vocabulary support and concrete examples for the below-grade-level students.

How to Use This Product

Readability Chart

Title of the Text	Star	Circle	Square	Triangle
Angles All Around	2.2	3.2	5.0	6.6
Understanding Triangles	1.5	3.5	5.5	6.7
To Cross or Not to Cross	2.1	3.3	5.4	6.8
Quadrilaterals	2.1	3.0	5.1	6.5
Classifying 2-D Shapes	2.2	3.5	5.0	6.6
Irregular Shapes	1.8	3.5	5.5	6.8
Congruent and Similar Figures	2.2	3.4	5.0	6.5
Understanding 3-D Shapes	2.2	3.5	5.0	6.7
Understanding Prisms	2.0	3.1	5.0	6.6
The Coordinate Plane	1.8	3.4	5.4	6.7
Circles	2.0	3.0	5.1	6.5
Symmetry	1.9	3.1	5.1	6.7
Reflections	2.2	3.4	5.4	6.5
Rotations	2.2	3.0	5.5	7.0
Translations	2.2	3.5	5.4	6.8

Components of the Product

Primary Sources

- Each level of text includes multiple primary sources. These documents, photographs, and illustrations add interest to the texts. The mathematic images also serve as visual support for second language learners. They make the texts more context-rich and bring the examples to life.

How to Use This Product (cont.)

Components of the Product (cont.)

Practice Problems

- The introduction often includes a challenging question or riddle. The answer can be found on the next page at the end of the lesson.
- Each level of text includes a You Try It section where the students are asked to solve problems using the skill or concept discussed in the text.
- Although the mathematics is the same, the questions may be worded slightly differently depending on the reading level of the passage.

The Levels

- There are 15 topics in this book. Each topic is leveled to four different reading levels. The images and fonts used for each level within a topic look the same.
- Behind each page number, you'll see a shape. These shapes indicate the reading levels of each text so that you can make sure students are working with the correct texts. The reading levels fall into the ranges indicated below. See the chart on page 16 for the specific reading levels of each lesson.

Leveling Process

- The texts in this series were originally authored by mathematics educators. A reading expert went through the texts and leveled each one to create four distinct reading levels.
- A mathematics expert then reviewed each passage for accuracy and mathematical language.
- The texts were then leveled one final time to ensure the editorial changes made during the process kept them within the ranges described to the left

Levels 1.5–2.2

Levels 3.0–3.5

Levels 5.0–5.5

Levels 6.5–7.2

How to Use This Product (cont.)

Tips for Managing the Product

How to Prepare the Texts

- When you copy these texts, be sure you set your copier to copy photographs. Run a few test pages and adjust the contrast as necessary. If you want the students to be able to appreciate the images, you need to carefully prepare the texts for them.

- You also have full-color versions of the texts provided in PDF form on the CD. (See page 142 for more information.) Depending on how many copies you need to make, printing the full-color versions and copying those might work best for you.

- Keep in mind that you should copy two-sided to two-sided if you pull the pages out of the book. The shapes behind the page numbers will help you keep the pages organized as you prepare them.

Distributing the Texts

Some teachers wonder about how to hand the texts out within one classroom. They worry that students will feel insulted if they do not get the same papers as their neighbors. The first step in dealing with these texts is to set up your classroom as a place where all students learn at their individual instructional levels. Making this clear as a fact of life in your classroom is key. Otherwise, the students may constantly ask about why their work is different. You do not need to get into the technicalities of the reading levels. Just state it as a fact that every student will not be working on the same assignment every day. If you do this, then passing out the varied levels is not a problem. Just pass them to the correct students as you circle the room.

If you would rather not have students openly aware of the differences in the texts, you can try these ways to pass out the materials:

- Make a pile in your hands from star to triangle. Put your finger between the circle and square levels. As you approach each student, you pull from the top (star), above your finger (circle), below your finger (square), or the bottom (triangle). If you do not hesitate too much in front of each desk, the students will probably not notice.

- Begin the class period with an opening activity. Put the texts in different places around the room. As students work quietly, circulate and direct students to the right locations for retrieving the texts you want them to use.

- Organize the texts in small piles by seating arrangement so that when you arrive at a group of desks you have just the levels you need.

How to Use This Product (cont.)

Correlation to Mathematics Standards

Shell Education is committed to producing educational materials that are research and standards based. In this effort, we have correlated all of our products to the academic standards of all 50 United States, the District of Columbia, the Department of Defense Dependent Schools, and all Canadian provinces. We have also correlated to the Common Core State Standards.

How to Find Standards Correlations

To print a customized correlation report of this product for your state, visit our website at **http://www.shelleducation.com** and follow the on-screen directions. If you require assistance in printing correlation reports, please contact Customer Service at 1-877-777-3450.

Purpose and Intent of Standards

Legislation mandates that all states adopt academic standards that identify the skills students will learn in kindergarten through grade twelve. Many states also have standards for Pre-K. This same legislation sets requirements to ensure the standards are detailed and comprehensive.

Standards are designed to focus instruction and guide adoption of curricula. Standards are statements that describe the criteria necessary for students to meet specific academic goals. They define the knowledge, skills, and content students should acquire at each level. Standards are also used to develop standardized tests to evaluate students' academic progress. Teachers are required to demonstrate how their lessons meet state standards. State standards are used in the development of all of our products, so educators can be assured they meet the academic requirements of each state.

TESOL Standards

The lessons in this book promote English language development for English language learners. The standards listed on the Teacher Resource CD support the language objectives presented throughout the lessons.

NCTM Standards Correlation Chart

The chart on the next page shows the correlation to the National Council for Teachers of Mathematics (NCTM) standards. This chart is also available on the Teacher Resource CD (*nctm.pdf*).

NCTM Standards

NCTM Standard	Lesson	Page(s)
Identifies, compares and analyzes attributes of two- and three-dimensional shapes and develops vocabulary to describe the attributes	Angles All Around; To Cross or Not to Cross; Quadrilaterals; Classifying 2-D Shapes; Understanding 3-D Shapes; Understanding Prisms; Circles	21–28, 37–60, 77–92, 101–108
Classifies two- and three-dimensional shapes according to their properties and develops definitions of classes of shapes such as triangles and pyramids	Understanding Triangles; Quadrilaterals; Classifying 2-D Shapes	29–36, 45–60
Investigates, describes, and reasons about the results of subdividing, combining, and transforming shapes	Irregular Shapes	61–68
Explores congruence and similarity	Understanding Triangles; Quadrilaterals; Classifying 2-D Shapes; Irregular Shapes; Congruent and Similar Figures; Translations	29–36, 45–76, 133–140
Describes location and movement using common language and geometric vocabulary	The Coordinate Plane	93–100
Makes and uses coordinate systems to specify locations and to describe paths	The Coordinate Plane	93–100
Finds the distance between points along horizontal and vertical lines of a coordinate system	The Coordinate Plane	93–100
Predicts and describes the results of sliding, flipping, and turning two-dimensional shapes	Reflections; Rotations; Translations	117–140
Describes a motion or a series of motions that will show that two shapes are congruent	Reflections; Rotations; Translations	117–140
Identifies and describes line and rotational symmetry in two- and three-dimensional shapes and designs	Symmetry	109–116
Builds and draws geometric objects	Congruent and Similar Figures; Understanding Prisms; Reflections; Rotations; Translations	69–76, 85–92, 117–140
Creates and describes mental images of objects, patterns and paths	Congruent and Similar Figures; Symmetry	69–76, 109–116
Identifies and builds a three-dimensional object from two-dimensional representations of that object	Understanding 3-D Shapes	77–84
Identifies and draws a two-dimensional representation of a three-dimensional object	Understanding 3-D Shapes	77–84
Uses geometric models to solve problems in other areas of mathematics, such as number and measurement	Irregular Shapes	61–68
Recognizes geometric ideas and relationships and applies them to other disciplines and to problems that arise in the classroom or in everyday life	Irregular Shapes; Reflections; Rotations	61–68, 117–132

Standards are listed with the permission of the National Council of Teachers of Mathematics (NCTM). NCTM does not endorse the content or validity of these alignments.

Angles All Around

Do you like water parks? Angles are used to design the slides.

Basic Facts

All figures are made up of **points**. A **ray** starts at one point. It stretches in one direction. It goes on forever. An **angle** has two rays. They intersect at one point. That point is called the **vertex**.

- Point
- Ray
- Vertex — Angle

Naming Angles

You know the word *angle*. We can use a symbol instead. The symbol ∠ can be read as "angle."

There are three ways to name angles. We can name them by number. We can name them by the vertex. We can name them with three points. One point is on one ray. One point is on the vertex. One point is on the other ray. The vertex must be the second point in the name (∠ABC).

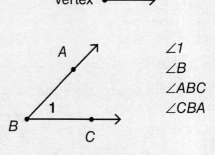

∠1
∠B
∠ABC
∠CBA

Measuring Angles

Angles are measured. We use **degrees**. We use a protractor to measure angles. Look at the protractor below. We read the measure of the angle with the outer numbers. One ray passes through zero. The other ray passes through 50. This angle measures 50°.

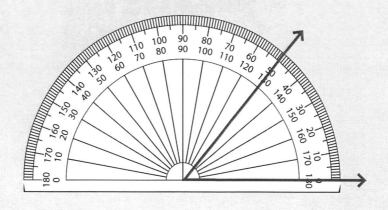

Types of Angles

There are many types of angles. We can group them. The groups are based on the measure of the angle.

Acute Angle

An **acute angle** has a measure that is less than 90°. ∠ABC and ∠JKL are acute angles. They are both less than 90°.

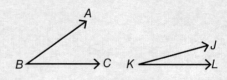

Right Angle

A **right angle** measures 90°. Angles E, F, D, and N all measure 90°, so they are all right angles. They are all right angles. There is a little box drawn at the vertex of angle D and angle N. That box shows that an angle is a right angle.

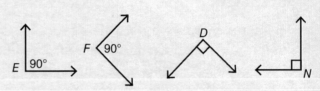

Obtuse Angle

An **obtuse angle** is greater than 90° but less than 180°. ∠H and ∠R both measure 150°. ∠S has measures 140°. They are all obtuse angles.

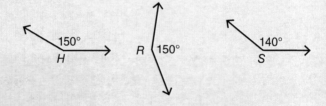

Angles in Our Daily Lives

It is important to understand angles. Think of roller coasters. The designers make sure the angles are not too steep. The roller coaster would be too dangerous. But the angles have to be steep enough to go fast. Riders want to be thrilled!

You Try It

Look at the figure below. Identify the angles. Are they acute, right, or obtuse?

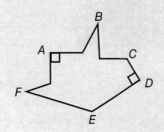

22

#50717—Leveled Texts for Mathematics: Geometry © Shell Education

Angles All Around

Have you ever been to a water park? Did you know that angles are used to design the slides?

Basic Facts

All figures are made up of **points**. A **ray** starts at one point. Then it stretches in one direction. It goes on forever. An **angle** is made up of two rays. They intersect at one point. This point where the two rays meet is called the **vertex**.

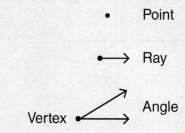

Naming Angles

You know the word *angle*. Sometimes we use a symbol instead. When you see the symbol ∠ you can read it as "angle."

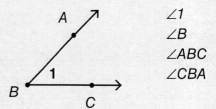

There are three ways to name angles. We can name them by number. We can name them by the vertex. We can name them with three points. If we use three points, they must have a point on one ray, a vertex, and a point on the other ray. The vertex must be the second point in the name (∠ABC).

Measuring Angles

Angles can be measured. They are measured in **degrees**. We use a protractor to measure angles. Look at the protractor below. We read the measure of the angle with the outer numbers. Notice that one ray passes through zero while the other passes through 50. That means that this angle measures 50°.

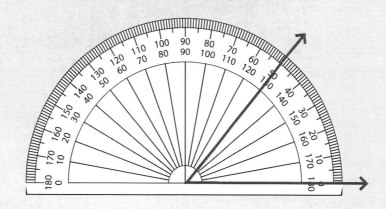

Types of Angles

There are many types of angles. They are based on the measure of the angle.

Acute Angle

An **acute angle** has a measure that is less than 90°. ∠ABC and ∠JKL are acute angles. They are both less than 90°.

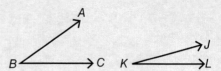

Right Angle

A **right angle** measures 90°. Angles E, F, D, and N all measure 90°, so they are all right angles. Did you notice the little box drawn at the vertex of angle D and angle N? That box shows that an angle is a right angle.

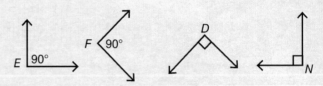

Obtuse Angle

An **obtuse angle** is greater than 90° but less than 180°. ∠H and ∠R both measure 150°. ∠S measures 140°. They are all obtuse angles.

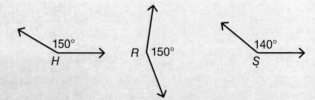

Angles in Our Daily Lives

Angles are important when building roller coasters. The designers have to make sure that the angles are not too steep. That might make the roller coaster too dangerous. But, the angles have to be steep enough to go fast. Riders want to be thrilled!

You Try It

Look at the figure below. Identify the acute angles, the right angles, and the obtuse angles.

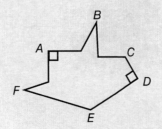

Angles All Around

Did you know that water park designers use angles to design water slides?

Basic Facts

All figures are made up of **points**. A **ray** starts at a single point and then stretches forever in one direction. An **angle** is made up of two rays that intersect at one point. This point where the two rays meet is the **vertex**.

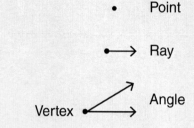

Naming Angles

Rather than write the word *angle* many times, we use a symbol. Whenever you see the symbol $\angle$ you can read it as "angle."

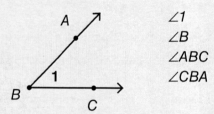

There are three ways to name angles: by number, by the vertex, or by three points. If we use three points, they must consist of a point on one ray, the vertex, and a point on the other ray. If you use three points to name an angle, the vertex must be the second point in the name ($\angle ABC$).

Measuring Angles

Angles are measured in **degrees**. To measure an angle, we use a protractor. Look at the protractor below. We read the measure of the angle with the outer numbers. Notice that one ray passes through zero while the other passes through 50, which means that this angle measures 50°.

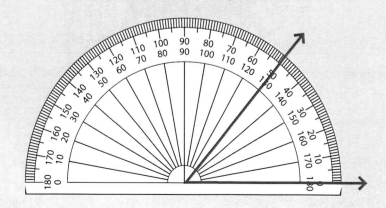

Types of Angles

There are different types of angles. These types are based on the measurement of the angle.

Acute Angle

An **acute angle** has a measure that is less than 90°. ∠ABC and ∠JKL are acute angles, because they are both less than 90°.

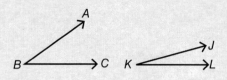

Right Angle

A **right angle** measures 90°. Angles E, F, D, and N all measure 90°, so they are all right angles. Did you notice the little box drawn at the vertex of angle D and angle N? That box shows that an angle is a right angle.

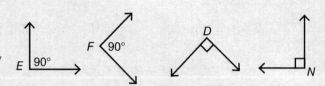

Obtuse Angle

An **obtuse angle** has a measure that is greater than 90° but less than 180°. ∠H and ∠R both have a measure of 150,° while ∠S has a measure of 140°. They are all obtuse angles.

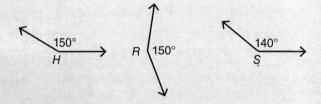

Angles in Our Daily Lives

Angles are important when building roller coasters. The designers have to make sure that the angles are not too steep, because that might make the roller coaster too dangerous. But, the angles have to be steep enough to go fast and make the ride exciting!

You Try It

Look at the figure below and identify the acute angles, the right angles, and the obtuse angles.

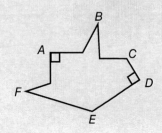

26

#50717—Leveled Texts for Mathematics: Geometry © Shell Education

Angles All Around

Did you know that water park designers use angles when they are designing the slides?

Basic Facts

All figures are made up of **points**. One example is a **ray**, which starts at a single point and then stretches forever in one direction. An **angle** is made up of two rays that intersect at a point, which is called the **vertex**.

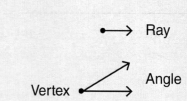

Naming Angles

Rather than writing the word *angle* repeatedly, we use a symbol to represent it. Whenever you see the symbol ∠ you can read it as "angle."

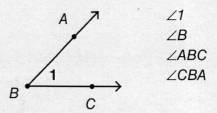

There are three ways to identify angles: by number, by the vertex, or by three points. If we use three points, they must consist of a point on one ray, the vertex, and a point on the other ray. If you use three points to name an angle, the vertex must be the second point in the name (∠ABC).

Measuring Angles

We use a protractor to measure the **degrees** of an angle. Look at the protractor below. We read the measure of the angle with the outer numbers. Notice that one ray passes through zero while the other passes through 50, which signifies that this angle measures 50°.

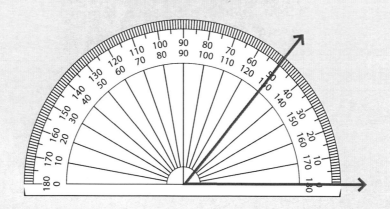

Types of Angles

Different types of angles can be identified by the measurement of the angle.

Acute Angle

An **acute angle** measures less than 90°. ∠ABC and ∠JKL are acute angles because they both measure less than 90°.

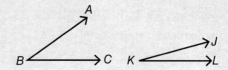

Right Angle

Every **right angle** measures 90°. Angles E, F, D, and N all measure 90°, so they are all right angles. Did you recognize the little box drawn at the vertex of angle D and angle N? That box identifies the angle as a right angle.

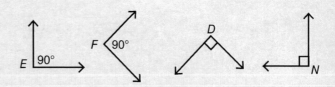

Obtuse Angle

An **obtuse angle** has a measure that is greater than 90° but less than 180°. ∠H and ∠R both have measurements of 150°, while ∠S measures 140°. They are all obtuse angles.

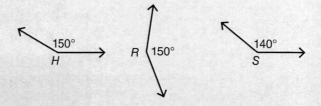

Angles in Our Daily Lives

Angles are critically important in the construction of roller coasters. Designers have to ensure that the angles are not too steep, because that might make the roller coaster too dangerous. However, the angles have to be steep enough to make for a thrilling ride!

You Try It

Look at the figure below and identify the acute angles, the right angles, and the obtuse angles.

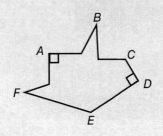

Understanding Triangles

Look at the design below. How many triangles do you see?

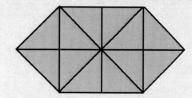

Basic Facts

A **triangle** is a closed shape. It is **two-dimensional**. It has three line segments for its sides. It is a three-sided polygon. The sides meet at three points. These points are called **vertices**. Each vertex forms an angle with two of the sides. The word *triangle* means "three angles." You can measure the three angles. You can add up the measures of those angles. They always add up to 180°.

There is a symbol used for triangles. It looks like this: ∆. There are letters by each vertex. You need to use these, too. This triangle can be identified six ways. They are: ∆LMN, ∆LNM, ∆MLN, ∆MNL, ∆NML, or ∆NLM.

How to Name Triangles

We can name triangles in two ways. They can be named by their angles. They can be named by their sides. We need to know this so that we can know how triangles are different.

Right Triangle

A right angle measures 90°. If a triangle has a right angle, it is called a **right triangle**. A triangle cannot have more than one right angle. There is a way to know if an angle is a right angle. We put a little box in the corner of the right angle.

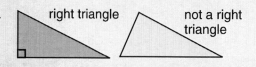

Acute Triangle

Some triangles have three acute angles. These are called **acute triangles**. Each angle is less than 90°.

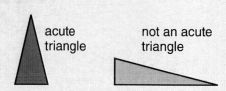

Obtuse Triangle

Some triangles have one obtuse angle. These are called **obtuse triangles**. One angle is greater than 90°. A triangle cannot have more than one obtuse angle.

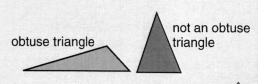

How to Name Triangles (cont.)

Equilateral Triangle

Some triangles have three sides that are the same. The sides are equal. The angles are equal, too. Each angle is 60°. These triangles are called **equilateral triangles**.

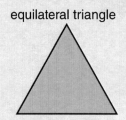

Isosceles Triangle

Some triangles have two equal sides. These are called **isosceles triangles**. They also have two equal angles.

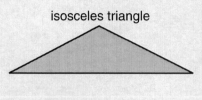

Scalene Triangle

Some triangles have no sides that are the same. They have no angles that are the same. These triangles are called **scalene triangles**.

Triangles in Our Daily Lives

Triangles can be used to build things. They are strong shapes. Some rooftops are shaped like triangles. So are some parts of ceilings and bridges.

Finding Missing Angles

You can find the measure of an angle in a triangle. Look at the triangle at right. First, find the measure of the unknown angle. All triangles have three angles. They add up to 180°.

Step 1: Add the two angles.

$$25° + 25° = 50° \text{ (sum)}$$

Step 2: Subtract the sum (of the two angles) from 180°.

$$180° - 50° = 130° \text{ (answer)}$$

$$\angle x = 130°$$

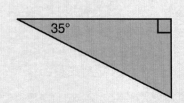

You Try It

This is a scalene triangle. It is also a right triangle. What is the missing angle in this triangle? **Hint:** The two known angles are 35° and 90°.

(28 triangles)

Understanding Triangles

Look at the design below. How many triangles do you see?

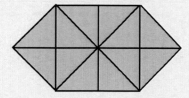

Basic Facts

A **triangle** is a closed shape. It is **two-dimensional**. It has three line segments for its sides. It is a three-sided polygon. The sides meet at three points. These points are called **vertices**. Each vertex forms an angle with two of the sides. The word *triangle* means "three angles." You can measure the three angles. You can add up the measurements of those angles. They always add up to 180°.

There is a symbol used for triangles. It looks like this: ∆. There are letters by each vertex. You need to use these, too. This triangle can be identified six ways. They are: ∆*LMN*, ∆*LNM*, ∆*MLN*, ∆*MNL*, ∆*NML*, or ∆*NLM*.

How to Name Triangles

We can name triangles in two ways. They can be named by their angles. They can be named by their sides. We need to know this so that we can know how triangles are different.

Right Triangle

A right angle measures 90°. If a triangle has a right angle, it is called a **right triangle**. A triangle cannot have more than one right angle. There is a way to know if an angle is a right angle. We put a little box in the corner of the right angle.

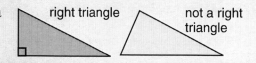

Acute Triangle

Some triangles have three acute angles. These are called **acute triangles**. Each angle is less than 90°.

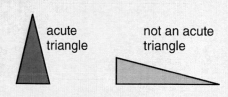

Obtuse Triangle

Some triangles have one obtuse angle. These are called **obtuse triangles**. One angle is greater than 90°. A triangle cannot have more than one obtuse angle.

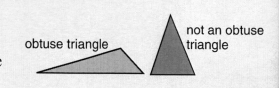

How to Name Triangles (cont.)

Equilateral Triangle

Some triangles have three sides that are the same. The sides are equal. The angles are equal, too. Each angle is 60°. These triangles are called **equilateral triangles**.

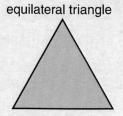
equilateral triangle

Isosceles Triangle

Some triangles have two equal sides. These are called **isosceles triangles**. They also have two equal angles.

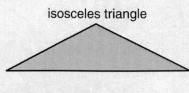
isosceles triangle

Scalene Triangle

Some triangles have no sides that are the same. They have no angles that are the same. These triangles are called **scalene triangles**.

scalene triangle

Triangles in Our Daily Lives

Triangles can be used to build things. They are strong shapes. Some rooftops are shaped like triangles. So are parts of some ceilings and bridges.

Finding Missing Angles

You can find the measure of an angle in a triangle. Look at the triangle at right. First, find the measure of the unknown angle. All triangles have three angles. The three angles add up to 180°.

Step 1: Add the two known angles.

$$25° + 25° = 50° \text{ (sum)}$$

Step 2: Subtract the sum (of the two angles) from 180°.

$$180° - 50° = 130° \text{ (answer)}$$

$$\angle x = 130°$$

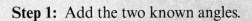

You Try It

This is a scalene triangle. It is also a right triangle. What is the missing angle in this triangle? **Hint:** The two known angles are 35° and 90°.

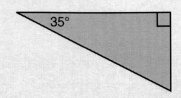

#50717—Leveled Texts for Mathematics: Geometry © Shell Education

Understanding Triangles

Look at the figure below. How many triangles do you see?

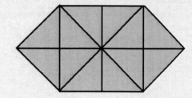

Basic Facts

A **triangle** is a closed **two-dimensional** figure with three line segments for its sides. It is a three-sided polygon. The sides meet at three points called **vertices**. Each vertex forms an angle with two of the sides. The word *triangle* means "three angles." If you add up the measures of the three angles of a triangle, they will always total 180°.

There is a specific symbol used to identify triangles. It looks like this: Δ. You use this symbol with the letters by each vertex. So, this triangle can be identified in the following ways: ΔLMN, ΔLNM, ΔMLN, ΔMNL, ΔNML, or ΔNLM.

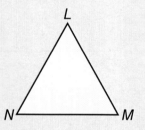

How to Name Triangles

There are two ways to name triangles: by their angles, or by their sides. This is important to know. It helps explain the differences among triangles.

Right Triangle

A right angle measures exactly 90°. If a triangle has a right angle, it is called a **right triangle**. A triangle cannot have more than one right angle. So that people know an angle is a right angle, you need to put a little box in the corner of the right angle.

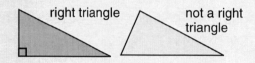

Acute Triangle

If a triangle has three acute angles, it is called an **acute triangle**. Each angle is less than 90°.

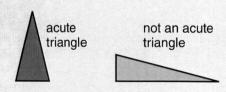

Obtuse Triangle

If a triangle has one obtuse angle, it is called an **obtuse triangle**. An obtuse angle is greater than 90°. A triangle cannot have more than one obtuse angle.

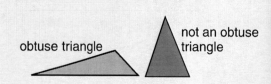

How to Name Triangles (cont.)

Equilateral Triangle

An **equilateral triangle** has three equal sides and three equal angles. All three angles in an equilateral triangle are 60°.

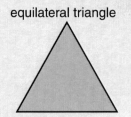
equilateral triangle

Isosceles Triangle

An **isosceles triangle** has only two equal sides. This special kind of triangle also has two equal angles. The angles opposite the equal sides are the same measure.

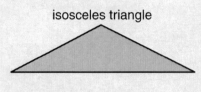
isosceles triangle

Scalene Triangle

A **scalene triangle** has no equal sides. These triangles also have no equal angles.

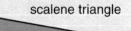
scalene triangle

Triangles in Our Daily Lives

Triangles can be used in architecture. They are strong shapes. You see them used in rooftops. You see them used in parts of floors and ceilings. You even see them used in bridges!

Finding Missing Angles

Sometimes, you have to figure out the measure of an angle in a triangle. You can do this if you know the measures of the other two angles in the triangle. Every triangle has three angles that always add up to 180°. So, you can use addition and then subtraction to find any unknown angle. Look at the triangle at right.

Step 1: Add the known angles.

$25° + 25° = 50°$ (sum)

Step 2: Subtract the sum (of the known angles) from 180°.

$180° - 50° = 130°$ (answer)

$\angle x = 130°$

You Try It

Find the missing angle in this scalene right triangle.
Hint: The right angle is 90°.

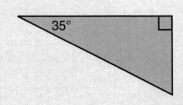

Understanding Triangles

How many triangles do you see in the design below?

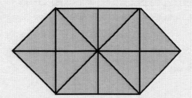

Basic Facts

A **triangle** is a closed **two-dimensional** figure with three line segments for its sides. It is a three-sided polygon. The sides meet at three points called **vertices**, each of which forms an angle with two of the sides. The word *triangle* means "three angles." If you add up the measurements of a triangle's three angles, they will always total 180°.

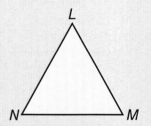

This is the specific symbol used to identify triangles: ∆. You also have to use the letters adjacent to each vertex. So, this triangle can be identified in the following six ways: ∆*LMN*, ∆*LNM*, ∆*MLN*, ∆*MNL*, ∆*NML*, or ∆*NLM*.

How to Name Triangles

There are two different methods to name triangles: by their angles, or by their sides. This is important to understand so that you can explain the differences among triangles.

Right Triangle

A right angle measures exactly 90°. If a triangle has a right angle, it is identified as a **right triangle**. It is impossible for a triangle to have more than one right angle. So that people can identify a right angle, you need to put a little box in the corner of the right angle.

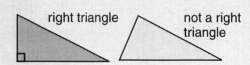

Acute Triangle

If a triangle has three acute angles, it is called an **acute triangle**. Each angle is less than 90°.

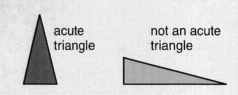

Obtuse Triangle

If a triangle has one obtuse angle, it is called an **obtuse triangle**. An obtuse angle is greater than 90°. It is impossible for a triangle to have more than one obtuse angle.

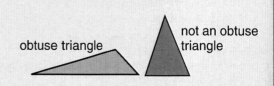

How to Name Triangles (cont.)

Equilateral Triangle

An **equilateral triangle** has three equal sides and three equal angles. Each angle is 60°.

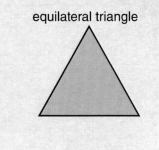
equilateral triangle

Isosceles Triangle

A triangle with two equal sides is called an **isosceles triangle**. Isosceles triangles also have two equal angles. The angles opposite the equal sides are the same measure.

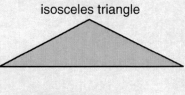
isosceles triangle

Scalene Triangle

Scalene triangles have no sides or angles that are the same.

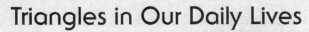
scalene triangle

Triangles in Our Daily Lives

Triangles are strong shapes and therefore, are used in various construction projects. Some parts of rooftops, ceilings, and even bridges are shaped like triangles.

Finding Missing Angles

Sometimes, you have to figure out the measure of an angle in a triangle. You can do this if you know the measures of the other two angles in the triangle. Every triangle has three angles that always add up to 180°. So, you can use addition and then subtraction to find any unknown angle. Look at the triangle at right.

Step 1: Add the two known angles.

$$25° + 25° = 50° \text{ (sum)}$$

Step 2: Subtract the sum (of the known angles) from 180°.

$$180° - 50° = 130° \text{ (answer)}$$

$$\angle x = 130°$$

You Try It

Find the missing angle in this scalene right triangle. **Hint:** The right angle is 90°.

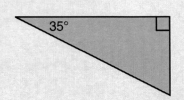

(28 triangles)

#50717—Leveled Texts for Mathematics: Geometry © Shell Education

To Cross or Not to Cross

Look at the train tracks. It may look like the tracks will meet. But the two steel rods will never intersect!

Basic Facts

Lines

Lines are endlessly straight. They go on and on. We draw arrows at each end of a line. This shows that it continues forever.

segment $\overline{AB}$ or line $\overleftrightarrow{AB}$

Parallel or Intersecting

There are many kinds of lines. One type of line is called **parallel**. Parallel lines will never cross. They will always be the same distance apart. When lines cross they **intersect** at one point.

parallel

intersecting

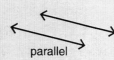
parallel

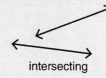
intersecting

A **ray** is a line with a starting point at one side. But it has no ending point. A **segment** is a line with two endpoints. Rays and segments can be parallel. They can also intersect.

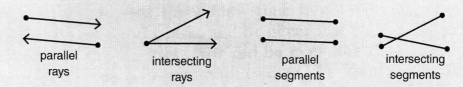

parallel rays · intersecting rays · parallel segments · intersecting segments

Perpendicular

Lines, rays, and segments can intersect. When they do, angles are formed. You know that a right angle has a measure of 90°. When right angles are formed, the lines, rays, or segments that formed those angles are called **perpendicular**.

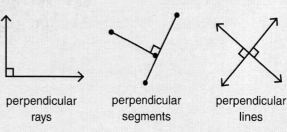

perpendicular rays · perpendicular segments · perpendicular lines

Working with Parallel and Perpendicular Lines

Let's find all the pairs of parallel and perpendicular sides in the examples below.

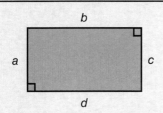

Parallel
a and c
b and d

Perpendicular
a and d   d and c
c and b   a and b

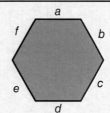

Parallel
a and d
b and e
c and f

Perpendicular
none

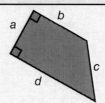

Parallel
b and d

Perpendicular
a and d
a and b

Parallel and Perpendicular Lines in Our Daily Lives

Look at the room shown. There are many examples of both types of lines. The pictures, table, chair, and drawers are only a few. If the handrail was not parallel with the base of the posts, it could cause problems. If the stair steps were not parallel with each other and with the ground, then anyone using the steps would likely fall!

You Try It

Find each pair of parallel and perpendicular sides in the figure below.

Parallel
f and ___
f and ___
a and ___
e and ___
d and ___

Perpendicular
d and ___
g and ___
g and ___

To Cross or Not to Cross

Look at the train tracks. Even though it may look like it, the two steel rods will never intersect!

Basic Facts

Lines

Lines are endlessly straight. They go on and on. We draw arrows at each end of a line. This shows that it continues forever.

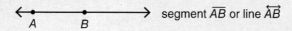

Parallel or Intersecting

There are many kinds of lines. One type of line is called **parallel**. Parallel lines will never cross. They will always be the same distance apart. If lines were to cross they would **intersect** at one point. When lines cross each other we say that they intersect.

  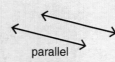 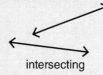

A **ray** is a line that has a starting point. But it has no ending point. A **segment** is a line with two endpoints. Rays and segments can be parallel. They can also intersect.

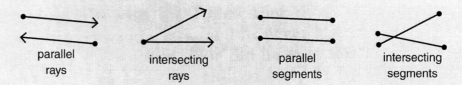

Perpendicular

Lines, rays, and segments can intersect. When they do, angles are formed. You know that a right angle has a measure of 90°. When right angles are formed, the lines, rays, or segments that formed those angles are called **perpendicular**.

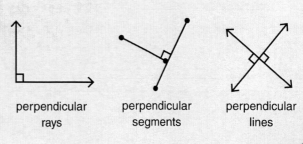

Working with Parallel and Perpendicular Lines

Let's find all the pairs of parallel and perpendicular sides in the examples below.

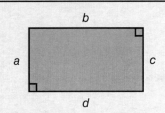

Parallel
a and c
b and d

Perpendicular
a and d   d and c
c and b   a and b

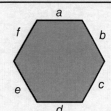

Parallel
a and d
b and e
c and f

Perpendicular
none

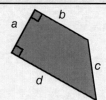

Parallel
b and d

Perpendicular
a and d
a and b

Parallel and Perpendicular Lines in Our Daily Lives

Look at the room shown. There are many examples of both types of lines. The pictures, table, chair, and drawers are only a few. If the handrail was not parallel with the base of the posts, it could cause problems. If the stair steps were not parallel with each other and with the ground, then anyone using the steps would likely fall!

You Try It

Find each pair of parallel and perpendicular sides in the figure below.

Parallel
f and ___
f and ___
a and ___
e and ___
d and ___

Perpendicular
d and ___
g and ___
g and ___

To Cross or Not to Cross

Look at the train tracks. Even though it may look like it, the two steel rods will never intersect!

Basic Facts

Lines

Lines are endlessly straight and endlessly long. We draw arrows at each end of a line to show that it continues forever.

segment $\overline{AB}$ or line $\overleftrightarrow{AB}$

Parallel or Intersecting

There are many different kinds of lines. One type of line is called **parallel**. Parallel lines will never cross and will always be the same distance apart. If lines were to cross, they would **intersect** at one point. Whenever lines cross one another we say that they are intersecting lines.

  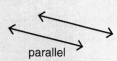 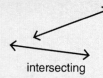

A **ray** is a line that has a starting point, but no ending point, and a **segment** is a line with two endpoints. Rays and segments can be parallel or they can intersect.

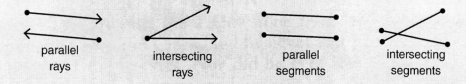

Perpendicular

When lines, rays, or segments intersect, angles are formed. Remember that a right angle has a measure of 90°. When right angles are formed, the lines, rays, or segments that formed those angles are called **perpendicular**.

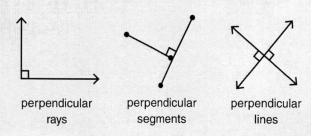

Working with Parallel and Perpendicular Lines

Let's find all the pairs of parallel and perpendicular sides in the examples below.

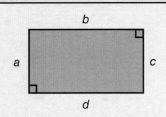 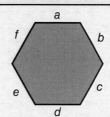 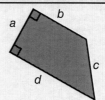

Parallel
a and c
b and d

Parallel
a and d
b and e
c and f

Parallel
b and d

Perpendicular
a and d   d and c
c and b   a and b

Perpendicular
none

Perpendicular
a and d
a and b

Parallel and Perpendicular Lines in Our Daily Lives

Look at the room shown here. There are many examples of both types of lines. The pictures, table, chair, and drawers are only a few. If the handrail was not parallel with the base of the posts, it could cause problems, and if the stair steps were not parallel with each other and with the ground, then anyone using the steps would surely be at risk of falling!

You Try It

Find each pair of parallel and perpendicular sides in the figure below.

Parallel
f and ___
f and ___
a and ___
e and ___
d and ___

Perpendicular
d and ___
g and ___
g and ___

To Cross or Not to Cross

Even though it may look like it, the two steel rods of these train tracks will never intersect!

Basic Facts

Lines

Lines are endlessly straight and endlessly long. We draw arrows at each end of a line to show that it continues indefinitely.

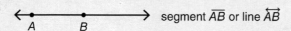

Parallel or Intersecting

There are various types of lines, including **parallel** and **intersecting** lines. Parallel lines will never cross and will always be the same distance apart. Whenever lines cross one another we call them intersecting lines.

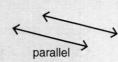
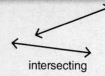

A **ray** is a line that has a starting point, but no ending point, and a **segment** is a line with two endpoints. Rays and segments can be parallel or they can intersect.

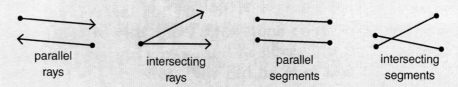

Perpendicular

When lines, rays, or segments intersect, angles are formed. Remember that a right angle has a measure of 90°. When right angles are formed, the lines, rays, or segments that formed those angles are called **perpendicular**.

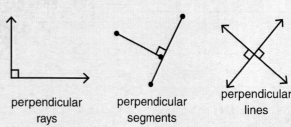

Working with Parallel and Perpendicular Lines

Let's find all the pairs of parallel and perpendicular sides in the examples below.

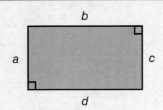 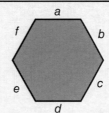 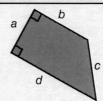

Parallel
 a and c
 b and d

Parallel
 a and d
 b and e
 c and f

Parallel
 b and d

Perpendicular
 a and d   d and c
 c and b   a and b

Perpendicular
 none

Perpendicular
 a and d
 a and b

Parallel and Perpendicular Lines in Our Daily Lives

Study the room pictured here, and you can find many examples of both types of lines. The pictures, table, chair, and drawers are only a few. If the handrail was not parallel with the base of the posts, it could cause problems, and if the stair steps were not parallel with each other and with the ground, then anyone using the steps would surely be at risk of falling!

You Try It

Find each pair of parallel and perpendicular sides in the figure below.

Parallel		**Perpendicular**
f and ___		d and ___
f and ___		g and ___
a and ___		g and ___
e and ___		
d and ___		

44

#50717—Leveled Texts for Mathematics: Geometry   © Shell Education

Quadrilaterals

A square is drawn. A second square is drawn inside the first square. A third square is drawn inside the second square. There are many polygons in this design. Can you count them all?

Basic Facts

A **polygon** is a closed figure. It is two-dimensional. It is made up of segments. These are called the **sides**. When two sides meet at their endpoints, they form a vertex.

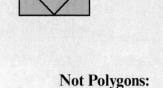

Polygons: **Not Polygons:**

A polygon with four sides is a **quadrilateral**. Quadrilaterals have four angles. They have four vertices, too. The sum of their angles is 360°.

Opposite Sides

The shapes below are all quadrilaterals. The first is a general quadrilateral. The other three are special ones.

Quadrilateral	**Parallelogram**
Opposite Sides: $\overline{EH}$ and $\overline{FG}$	Opposite Sides: $\overline{IJ}$ and $\overline{LK}$ — These two are parallel.
$\overline{EF}$ and $\overline{HG}$	$\overline{IL}$ and $\overline{JK}$ — These two are parallel.
Trapezoid	**Rhombus**
Opposite Sides: $\overline{RS}$ and $\overline{UT}$ — These two are parallel.	Opposite Sides: $\overline{MQ}$ and $\overline{NP}$ — These two are parallel.
$\overline{RU}$ and $\overline{ST}$ — These two are never parallel.	$\overline{MN}$ and $\overline{QP}$ — These two are parallel.

Opposite Angles

The shapes listed below are all quadrilaterals. The first is a **general quadrilateral**. The other two are special ones. Look at the angles inside each one. One pair of opposite angles is marked. One pair is not marked. This pair makes up the other pair of opposite angles.

Quadrilateral	Parallelogram	Rectangle
	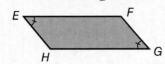	
Opposite Angles: ∠A and ∠C ∠B and ∠D	Opposite Angles: ∠E and ∠G ∠F and ∠H	Opposite Angles: ∠I and ∠K ∠L and ∠J All four angles are right angles.

Quadrilaterals in Our Daily Lives

People who design buildings are called architects. Their building plans are called blueprints. These drawings show the size and shape of the building. They show the supplies needed. Architects use these tools in their work. They help them draw quadrilaterals in their designs.

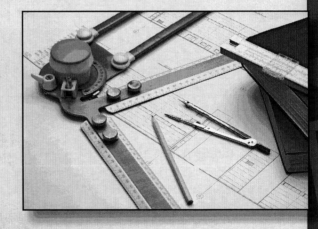

You Try It

The figure shown below is a trapezoid. Fill in the blanks with the correct answers.

Opposite Sides: $\overline{AX}$ and _____; $\overline{AM}$ and _____

Opposite Angles: ∠A and ∠ _____;

∠M and ∠ _____

Parallel Sides: _____ and _____

Congruent Angles: ∠ _____ and ∠ _____

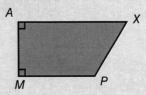

(11 polygons)

Quadrilaterals

A square is drawn. A second square is drawn inside the first square. A third square is drawn inside the second square. There are many polygons in this design. Can you count them all?

Basic Facts

A **polygon** is a closed figure. It is two-dimensional. It is made up of segments. These are called the **sides**. When two sides meet at their endpoints, they form a vertex.

Polygons: **Not Polygons:**

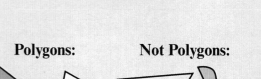

A polygon with four sides is a **quadrilateral**. Quadrilaterals have four angles, which means they also have four vertices. The sum of their angles is 360°.

Opposite Sides

The shapes below are all quadrilaterals. The first is a general quadrilateral. The other three are special ones.

Quadrilateral	Parallelogram
Opposite Sides:	Opposite Sides:
$\overline{EH}$ and $\overline{FG}$	$\overline{IJ}$ and $\overline{LK}$ These two are parallel.
$\overline{EF}$ and $\overline{HG}$	$\overline{IL}$ and $\overline{JK}$ These two are parallel.
Trapezoid	**Rhombus**
Opposite Sides:	Opposite Sides:
$\overline{RS}$ and $\overline{UT}$ These two are parallel.	$\overline{MQ}$ and $\overline{NP}$ These two are parallel.
$\overline{RU}$ and $\overline{ST}$ These two are never parallel.	$\overline{MN}$ and $\overline{QP}$ These two are parallel.

Opposite Angles

The shapes listed below are all quadrilaterals. The first is a general quadrilateral. The other two are special ones. Look at the angles in each one. One pair of opposite angles is marked but the other pair is not marked. This second pair represents the other pair of opposite angles.

Quadrilateral	Parallelogram	Rectangle
	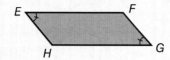	
Opposite Angles: $\angle A$ and $\angle C$ $\angle B$ and $\angle D$	Opposite Angles: $\angle E$ and $\angle G$ $\angle F$ and $\angle H$	Opposite Angles: $\angle I$ and $\angle K$ $\angle L$ and $\angle J$ All four angles are right angles.

Quadrilaterals in Our Daily Lives

People who design buildings are called architects. Their building plans are called blueprints. These drawings show the size and shape of the building. They show the supplies needed. Architects use these tools in their work. They help them draw quadrilaterals in their designs.

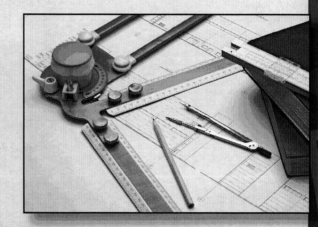

You Try It

The figure shown below is a trapezoid. Fill in the blanks with the correct answers.

Opposite Sides: $\overline{AX}$ and _____; $\overline{AM}$ and _____

Opposite Angles: $\angle A$ and $\angle$ _____;

$\angle M$ and $\angle$ _____

Parallel Sides: _____ and _____

Congruent Angles: $\angle$ _____ and $\angle$ _____

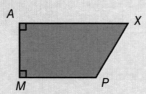

Quadrilaterals

Inside one drawn square is a second square. Inside the second square, a third square is drawn. How many polygons have been made in this design?

Basic Facts

A **polygon** is a closed, two-dimensional figure made up of segments that are called **sides**. When two sides meet at their endpoints, they form a vertex.

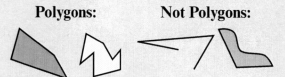

A polygon with four sides is a **quadrilateral**. Quadrilaterals also have four angles and four vertices. The sum of their angles equals 360°.

Opposite Sides

Some quadrilaterals are shown below. The first is a general quadrilateral and the other three are special types of quadrilaterals.

Quadrilateral	Parallelogram
Opposite Sides:	Opposite Sides:
$\overline{EH}$ and $\overline{FG}$	$\overline{IJ}$ and $\overline{LK}$ — These two are parallel.
$\overline{EF}$ and $\overline{HG}$	$\overline{IL}$ and $\overline{JK}$ — These two are parallel.
Trapezoid	**Rhombus**
Opposite Sides:	Opposite Sides:
$\overline{RS}$ and $\overline{UT}$ — These two are parallel.	$\overline{MQ}$ and $\overline{NP}$ — These two are parallel.
$\overline{RU}$ and $\overline{ST}$ — These two are never parallel.	$\overline{MN}$ and $\overline{QP}$ — These two are parallel.

Opposite Angles

The first figure below is a general quadrilateral and the others are special ones. In each case, one pair of opposite angles is marked. The remaining pair constitutes the other pair of opposite angles.

Quadrilateral	Parallelogram	Rectangle

Opposite Angles: ∠A and ∠C ∠B and ∠D	Opposite Angles: ∠E and ∠G ∠F and ∠H	Opposite Angles: ∠I and ∠K ∠L and ∠J All four angles are right angles.

Quadrilaterals in Our Daily Lives

Architects design blueprints of buildings before they are built. These two-dimensional drawings help architects know the size, shape, and materials they will need in order to bring the building to life. These tools help architects draw quadrilaterals on their blueprints when they are designing new buildings.

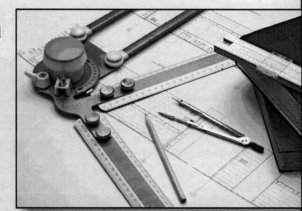

You Try It

The figure shown below is a trapezoid. Fill in the blanks with the correct answers.

Opposite Sides: $\overline{AX}$ and _____; $\overline{AM}$ and _____

Opposite Angles: ∠A and ∠ _____;

∠M and ∠ _____

Parallel Sides: _____ and _____

Congruent Angles: ∠ _____ and ∠ _____

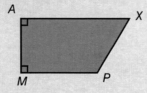

Quadrilaterals

There are three squares drawn here, one square inside the first square, and a third one inside the second. How many polygons can you count inside this design?

Basic Facts

A **polygon** is a closed, two-dimensional figure made up of segments that are called **sides**. When two sides meet at their endpoints, they form a vertex.

A polygon with four sides, four angles, and four vertices is a **quadrilateral**. The sum of its angles equals 360°.

Opposite Sides

The first of the quadrilaterals shown below is a general quadrilateral and the other three shapes are special types of quadrilaterals.

Quadrilateral	Parallelogram
Opposite Sides:	Opposite Sides:
$\overline{EH}$ and $\overline{FG}$	$\overline{IJ}$ and $\overline{LK}$ These two are parallel.
$\overline{EF}$ and $\overline{HG}$	$\overline{IL}$ and $\overline{JK}$ These two are parallel.
Trapezoid	**Rhombus**
Opposite Sides:	Opposite Sides:
$\overline{RS}$ and $\overline{UT}$ These two are parallel.	$\overline{MQ}$ and $\overline{NP}$ These two are parallel.
$\overline{RU}$ and $\overline{ST}$ These two are never parallel.	$\overline{MN}$ and $\overline{QP}$ These two are parallel.

Opposite Angles

The first of the quadrilaterals shown below is a general quadrilateral and the other two shapes are special types of quadrilaterals. In each case, one pair of opposite angles is marked, while the other pair of opposite angles is unmarked.

Quadrilateral	Parallelogram	Rectangle
	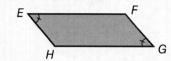	
Opposite Angles: $\angle A$ and $\angle C$ $\angle B$ and $\angle D$	Opposite Angles: $\angle E$ and $\angle G$ $\angle F$ and $\angle H$	Opposite Angles: $\angle I$ and $\angle K$ $\angle L$ and $\angle J$ All four angles are right angles.

Quadrilaterals in Our Daily Lives

Long before construction of a new building begins, an architect creates drawings called blueprints, that provide all measurements of the building and the materials required to build it. The tools shown assist architects in drawing quadrilaterals and other polygons on their blueprints when they are designing new buildings.

You Try It

The figure shown below is a trapezoid. Fill in the blanks with the correct answers.

Opposite Sides: $\overline{AX}$ and _____ ; $\overline{AM}$ and _____

Opposite Angles: $\angle A$ and $\angle$ _____ ;

$\angle M$ and $\angle$ _____

Parallel Sides: _____ and _____

Congruent Angles: $\angle$ _____ and $\angle$ _____

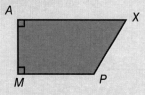

Classifying 2-D Shapes

How many rectangles do you see? What other shapes do you see?

Basic Facts

What does it mean to classify shapes? It means to sort the shape into a group or groups.

Do you see the nine rectangles in the shape person? The body and mouth are two. The arms and legs make four more. The feet and the chest are rectangles, too. They can be easily missed. They are also squares. A square is a type of rectangle. In fact, a square can have other labels, too. It is also a polygon. And it is a quadrilateral!

Classifying in Many Ways

Look at the nine figures below. Could you sort them into two groups? Could you sort them into three groups?

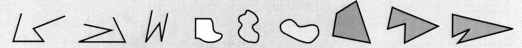

Sort them into two groups. One group could be "figures that do not close." The other group could be "figures that close."

Sort them into three groups. One group could be "figures that do not close." One group could be "closed figures with curves." The last group could be "closed figures with only segments."

The last three figures are polygons. A polygon is a closed two-dimensional figure made up only of segments. So, another way to group the figures is as "polygons" and "not polygons."

Can you think of other ways to group them?

Classifying Based on Sides

The number of sides that a polygon has can be used to classify the polygon.

Figure	Triangle	Quadrilateral	Pentagon	Hexagon	Heptagon	Octagon	Nonagon	Decagon
Number of Sides	3	4	5	6	7	8	9	10
Example	△	⏢	⬠	⬡	⬡	⯃	⬢	⯃

Classifying 2-D Shapes in Our Daily Lives

Look at the shapes in the design. There are squares, circles, and triangles. You know how to classify those shapes. You know what it means to be a square, circle, and triangle. When you know the type of figure you have, you can understand its properties. The artist who made this mosaic probably understood the properties of squares. He used what he knew to find the center of the circles.

You Try It

To what group could all four figures belong?

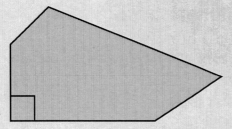
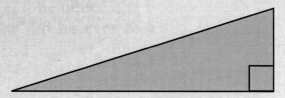
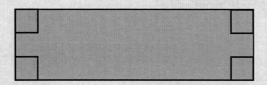
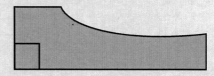

(6 rectangles; circles, ovals, squares, parallelograms, trapezoids)

Classifying 2-D Shapes

How many rectangles do you see? What other shapes do you see?

Basic Facts

What does it mean to classify shapes? It means to sort the shape into a group or groups.

Did you see that there are nine rectangles in the shape person? The body and mouth are two. The arms and legs make four more. It can be easy to miss that the feet and the chest are rectangles, too. This is because they are also squares. A square is a type of rectangle. In fact, a square can have other labels, too. It is also a polygon and a quadrilateral!

Classifying in Many Ways

Let's look at the nine figures below. Can you think of a way to split them into two groups? How about three groups?

Let's try putting them into two groups. We could group some as "figures that do not close." The rest could be in a group of "figures that close."

Let's try dividing them into three groups. We could group some as "figures that do not close." Some could be grouped as "closed figures with curves." The rest could be grouped as "closed figures with only segments."

The last three figures are polygons. A polygon is a closed two-dimensional figure made up only of segments. So, another way to group the figures is as "polygons" and "not polygons."

Can you think of any other way to group them?

Classifying Based on Sides

The number of sides that a polygon has can be used to classify the polygon.

Figure	Triangle	Quadrilateral	Pentagon	Hexagon	Heptagon	Octagon	Nonagon	Decagon
Number of Sides	3	4	5	6	7	8	9	10
Example	▲	⏢	⬟	⬢	⬡	⬣	⬢	⬣

Classifying 2-D Shapes in Our Daily Lives

Look at the shapes in the design. It has squares, circles, and triangles. You know how to classify those shapes. You know what it means to be a square, circle, and triangle. When you know the type of figure you have, you can understand the properties about that figure. The artist who made this mosaic might have used the properties of squares to find the center of the circles.

You Try It

What is a classification in which all four figures could belong?

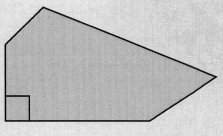
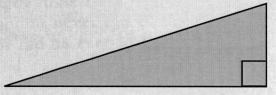
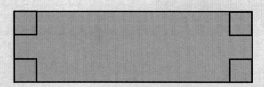
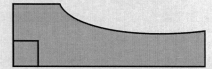

(6 rectangles; circles, ovals, squares, parallelograms, trapezoids)

Classifying 2-D Shapes

How many rectangles do you see? What are some of the other figures you see?

Basic Facts

When we classify shapes, we identify the group or groups to which a shape belongs.

Did you realize that there are nine rectangles in the shape person? The body and mouth are two. The arms and legs make four more. It can be easy to miss the feet and the chest as rectangles, because they are also squares. A square is simply a special type of rectangle. The square actually can have many labels. It is a polygon, a quadrilateral, a rectangle, and a square!

Classifying in Many Ways

Let's look at the nine figures below. Can you think of a way to split them into two groups? Is there a way to organize them into three groups?

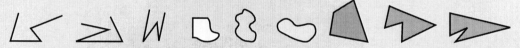

Let's try dividing the figures into two groups. We could group some as "figures that do not close" and others as "figures that close."

Let's try dividing the figures into three groups. We could group some as "figures that do not close," some as "closed figures with curves," and some as "closed figures with only segments."

The last three figures are polygons. A polygon is a closed two-dimensional figure made up only of segments. So, another way to group the figures is as "polygons" and "not polygons."

Is there another way that you might group the figures?

Classifying Based on Sides

The number of sides that a polygon has can be used to classify that polygon.

Figure	Triangle	Quadrilateral	Pentagon	Hexagon	Heptagon	Octagon	Nonagon	Decagon
Number of Sides	3	4	5	6	7	8	9	10
Example	△	⬢	⬟	⬢	⬣	⯃	⬣	⬣

Classifying 2-D Shapes in Our Daily Lives

Look at the squares, circles, and triangles in the mosaic. You know how to classify those shapes. You understand what it means to be a square, circle, and triangle. When you know the type of figure you have, you can understand the properties about that figure. The artist who made this mosaic might have used the properties of squares to find the center of the circles.

You Try It

Name a classification to which all four figures could belong.

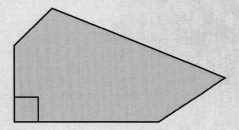
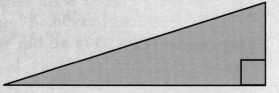
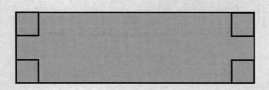
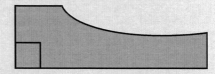

(6 rectangles; circles, ovals, squares, parallelograms, trapzoids)

#50717—Leveled Texts for Mathematics: Geometry © Shell Education

Classifying 2-D Shapes

How many rectangles can you find? Identify some of the other figures you see, too.

Basic Facts

To classify shapes, we identify the group or groups to which a shape belongs.

Did you realize that there are nine rectangles in the shape person? The body and mouth are two, and the arms and legs make four more. It can be easy to miss the feet and the chest as rectangles, because they are also squares. A square is simply a special type of rectangle. In fact, a square can be labeled a polygon, a quadrilateral, a rectangle, and a square!

Classifying in Many Ways

Look at the nine figures below and think of a way to split them into two groups. Is there also a way to organize them into three groups?

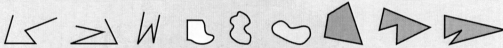

When dividing the figures into two groups, we could group some as "figures that do not close" and others as "figures that close."

When dividing the figures into three groups, we could group some as "figures that do not close," some as "closed figures with curves," and some as "closed figures with only segments."

The last three figures are polygons, which are closed, two-dimensional figures made up only of segments. So, another way to group the figures is as "polygons" and "not polygons."

Is there another way you can think of to organize the figures?

Classifying Based on Sides

The number of sides that a polygon has can be used to classify that polygon.

Figure	Triangle	Quadrilateral	Pentagon	Hexagon	Heptagon	Octagon	Nonagon	Decagon
Number of Sides	3	4	5	6	7	8	9	10
Example	△	▱	⬠	⬡	⬡	⯃	⬣	⬣

Classifying 2-D Shapes in Our Daily Lives

Look at the squares, circles, and triangles in the mosaic. You know how to classify those shapes because you understand the properties of squares, circles, and triangles. When you know the type of figure you have, you can understand the properties about that figure. The artist who made this mosaic might have used the properties of squares to find the center of the circles.

You Try It

Name a classification to which all four figures could belong.

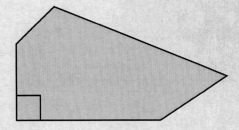
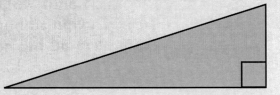
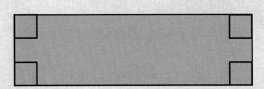
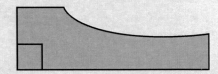

(6 rectangles; circles, ovals, squares, parallelograms, trapezoids)

Irregular Shapes

Look at the picture. See how many shapes you can find.

Basic Facts

A polygon is a closed shape. It is two-dimensional. It is made up only of segments. Polygons can be classified based on their number of sides. They can also be classified based on whether or not the sides all have the same length.

Regular Polygon

Some polygons are **regular**. This means that they are made up of sides that are all the same length. Their angles are all the same, too. These shapes are all regular polygons.

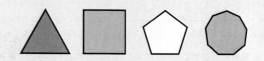

Irregular Polygon

All other polygons are **irregular**. Their sides and angles are not equal.

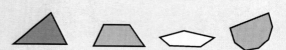

Polygons can be classified another way. It is based on whether all their sides somewhat "stick out" or if any "cave in."

Convex Polygon

All of the sides of a **convex polygon** seem to "stick out." Think of lines drawn along each side. Those lines would never go "inside" the polygon.

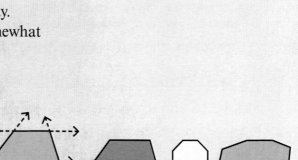

Concave Polygon

Some of the sides of a **concave polygon** seem to "cave in." Imagine lines drawn along each side that go "inside" the polygon.

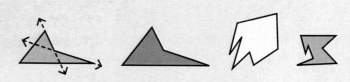

Look at the figures in the picture at the top of the page. Some are polygons. The sign is a regular polygon. It is convex. The flower is a regular polygon. It is concave. The sun is a regular polygon. It is convex.

Taking Shapes Apart

Have you noticed that many shapes can be broken up into simpler shapes?

Squares are regular polygons. They are convex. You can break them up. You can see two right triangles.

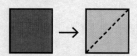

Trapezoids are irregular polygons. They are convex. You can break them up. You can see two triangles.

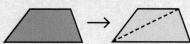

Some quadrilaterals are both irregular and concave. You can break them up. You can see two triangles.

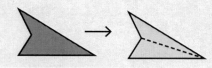

We can break up other polygons, too. They can be triangles, as well. Here is an irregular, concave polygon. It can be broken up into four triangles.

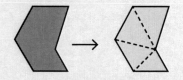

Irregular Shapes in Our Daily Lives

Think of a house that you have been in. What if it was all gone except for the floor? Would that floor be a rectangle? Maybe it would. But it most likely would not.

That floor covers a certain area. You could find the area if the floor were a rectangle. But what if the shape were irregular? We would have to break it up into pieces whose areas could be found more easily.

You Try It

Draw your own picture. Include three polygons. Use one that is convex and regular. Use one that is convex and irregular. Use one that is concave and irregular. List and describe your choices.

Irregular Shapes

Look at the picture. See how many shapes you can find.

Basic Facts

A polygon is a closed shape. It is two-dimensional. It is made up only of segments. Polygons can be classified based on their number of sides. They can also be classified based on whether or not the sides all have the same length.

Regular Polygon

Some polygons are **regular**. This means that they are made up of sides that are all the same length. Their angles are all the same, too. These shapes are all regular polygons.

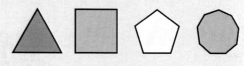

Irregular Polygon

All other polygons are **irregular**. Their sides and angles are not equal.

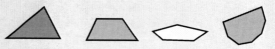

Polygons can be classified differently. Do their sides "stick out" or do any "cave in"?

Convex Polygon

All of the sides of a **convex polygon** seem to "stick out." Think of lines drawn along each side. Those lines would never go "inside" the polygon.

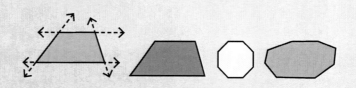

Concave Polygon

Some of the sides of a **concave polygon** seem to "cave in." Think of lines drawn along each side. Some of those lines would go "inside" the polygon.

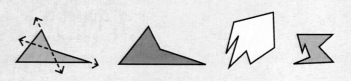

Look at the picture at the top again. Many figures are polygons. The sign is a regular polygon. It is convex. The flower is a regular polygon. It is concave. The sun is a regular polygon. It is convex.

Taking Shapes Apart

Have you noticed that many shapes can be broken up into simpler shapes?

Squares are regular polygons. They are convex. You can break them up. You can see two right triangles.

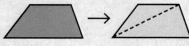

Trapezoids are irregular polygons. They are convex. You can break them up. You can see two triangles.

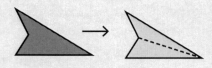

Some quadrilaterals are both irregular and concave. You can break them up. You can see two triangles.

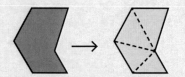

We can break up other polygons, too. They can be triangles, as well. Here is an irregular, concave polygon. It can be broken up into four triangles.

Irregular Shapes in Our Daily Lives

Think of a house that you have been in. What if it was all gone except for the floor? Would that floor be a rectangle? Maybe it would. But it most likely would not.

That floor covers a certain area. You could find the area if the floor were a rectangle. But what if the shape were irregular? We would have to break it up into pieces whose areas could be found more easily.

You Try It

Draw your own picture. Include three polygons. Use one that is convex and regular. Use one that is convex and irregular. Use one that is concave and irregular. List and describe your choices.

Irregular Shapes

Look at the picture to see how many different shapes you can find.

Basic Facts

A polygon is a closed two-dimensional figure. It is made up only of segments. Polygons are often classified based on their number of sides, but they can also be classified based on whether or not the sides all have the same length.

Regular Polygon

A **regular polygon** is made up of sides that are all the same length and angles that are all the same. The shapes here are all regular polygons.

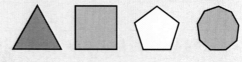

Irregular Polygon

Polygons that are not regular are **irregular**. Their sides and angles are not equal.

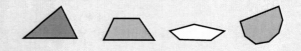

Polygons can also be classified based on whether all the sides somewhat "stick out" or if any "cave in."

Convex Polygon

All of the sides of a **convex polygon** seem to "stick out." If you were to draw lines along each side, those lines would never go "inside" the polygon.

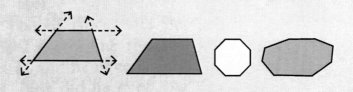

Concave Polygon

Some of the sides of a **concave polygon** seem to "cave in." If you were to draw lines along each side, some of those lines would go "inside" the polygon.

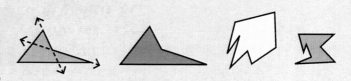

Look at the picture at the top again. Many figures are polygons. The sign is a convex, regular polygon. The flower is a concave, regular polygon. The sun is a convex, regular polygon.

Taking Shapes Apart

Have you noticed that many shapes can be broken up into simpler shapes?

Squares are regular, convex polygons. They can be broken up into two right triangles.

Trapezoids are irregular, convex polygons. They can be broken up into two triangles.

Some quadrilaterals are both irregular and concave. Even these can be broken up into two triangles.

Other polygons can be broken up into triangles as well. This irregular, concave polygon is broken up into four triangles.

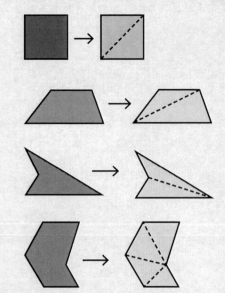

Irregular Shapes in Our Daily Lives

Think of a house that you have been in. Imagine taking all of it away except for the floor. Would that floor be a rectangle? Maybe it would, but probably not.

That floor covers a certain amount of area. You could probably easily find the area of the floor if it were a rectangle. But if it were an irregularly shaped floor, we would have to break it up into pieces whose areas could be found more easily.

You Try It

Draw your own picture. Include three polygons. Use one that is convex and regular. Use one that is convex and irregular. Use one that is concave and irregular. List and describe your choices.

Irregular Shapes

Examine the picture to see how many different shapes you can find.

Basic Facts

A polygon is a closed, two-dimensional figure consisting only of segments. Polygons are frequently classified based on their number of sides, but they can also be classified based on whether or not the sides all have the same length.

Regular Polygon

A **regular polygon** is made up of sides that are all equal in length and angles that are all equal. The shapes here are all regular polygons.

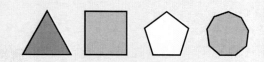

Irregular Polygon

Polygons that are not regular are classified as **irregular**. Their sides and angles are not equal.

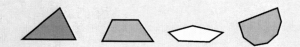

Polygons can also be classified based on whether all their sides somewhat "stick out" or if any "cave in."

Convex Polygon

All of the sides of **convex polygons** seem to "stick out." If you were to draw lines along each side, those lines would never go "inside" the polygon.

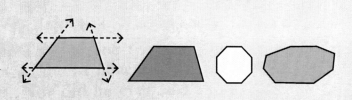

Concave Polygon

Some of the sides of **concave polygons** seem to "cave in." If you were to draw lines along each side, some of those lines would go "inside" the polygon.

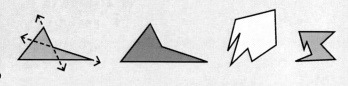

Look more closely at the picture at the top of the page again. Did you identify the figures as polygons? The sign is a convex, regular polygon, and the flower is a concave, regular polygon. The sun is a convex, regular polygon.

Taking Shapes Apart

Have you recognized that many shapes can be broken up into simpler shapes?

Squares are regular, convex polygons that can be broken up into two right triangles.

Trapezoids are irregular, convex polygons that can be broken up into two triangles.

Some quadrilaterals are both irregular and concave. Even these can be broken up into two triangles.

Other polygons can be broken up into triangles as well. This irregular, concave polygon is broken up into four triangles.

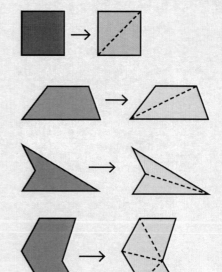

Irregular Shapes in Our Daily Lives

Think of a house that you have been in. Imagine taking all of it away except for the floor. Would that floor be a rectangle? Maybe it would, but probably not.

That floor covers a certain amount of area. You could probably easily find the area of the floor if it were a rectangle. But if it were an irregularly shaped floor, we would have to break it up into pieces whose areas could be found more easily.

You Try It

Draw your own picture that includes three polygons. Draw one that is convex and regular, one that is convex and irregular, and one that is concave and irregular. List and describe your choices.

Congruent and Similar Figures

Look at the triangles below. Which one does not belong?

Basic Facts of Congruence

Congruent segments have the same length.

You could place one on top of the other. There would be no difference between the two.

Congruent angles have the same measure.

You could place one on top of the other. That would be no difference between the two.

Two figures are congruent if they have the same size and shape. If you placed one on top of the other, there would be no difference between them.

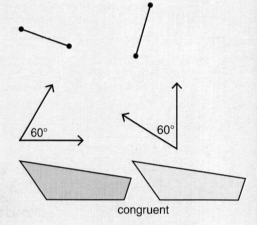
congruent

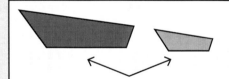
These figures are not congruent. They have different sizes.

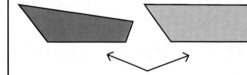
These figures are not congruent. They have different shapes.

Basic Facts of Similarity

Similar figures have the same shape. The sizes may be the same. The sizes may be different. Look at the similar figures. Look at the matching angles. The angles are congruent.

Figures *A*, *B*, and *C* are similar figures. Figure *D* has a different shape than the other three. That means that figures *A*, *B*, *C*, and *D* are not similar. There is a way to see that figures *A*, *B*, and *C* are similar. Imagine turning figure *C*. Turn it until it is in the same position as figures *A* and *B*. Figures *A* and *C* are also congruent.

Basic Facts of Similarity (cont.)

Look at the triangles at the top of the last page. They are not the same sizes. They have the same shape. All except one. Look at triangles A, B, C, and D. They are all similar. Triangle E does not belong.

Finding Missing Measurements

The two figures shown here are similar. We are going to see how to find the missing angle and side.

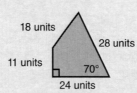
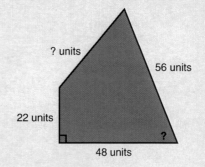

Angles: Matching angles of similar figures are congruent. So, the missing angle is 70°.

Sides: Each side of the first shape can be multiplied or divided by the same number to find each matching side of the second figure. For every side we know, we can multiply by 2 to find each matching side of the second figure. For example, $11 \times 2 = 22$. $28 \times 2 = 56$. And $24 \times 2 = 48$. So, the missing side is $18 \times 2 = 36$ units.

Similarity in Our Daily Lives

Look at the floor plan. This is a scale drawing of a house. Are these plans as large as a real house? Of course not! But, every angle is congruent with every matching angle of the real thing. We can find the length of every segment in the drawing. Then we can multiply each segment by the same number to find the length of the real thing. Scale drawings are used to plan homes and cars. They can even be used to plan 100-story buildings!

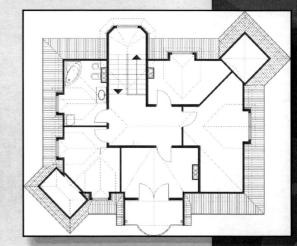

You Try It

The two triangles shown here are similar. Find the measure of the missing angle. Then find the length of the missing side. Now, draw a third triangle that is congruent with the first triangle.

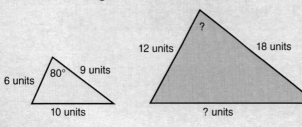

Congruent and Similar Figures

Look at the triangles. Which one does not belong?

Basic Facts of Congruence

Congruent segments have the same length.

You could lay one on top of the other. You would find that there is no difference in the two.

Congruent angles have the same measure.

You could lay one on top of the other. You would find that there is no difference in the two.

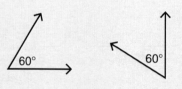

Two figures are congruent if they have the same size and shape. You could lay one on top of the other to find that there is no difference.

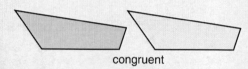
congruent

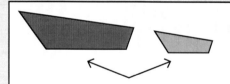
These figures are not congruent. They have different sizes.

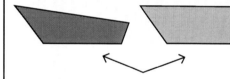
These figures are not congruent. They have different shapes.

Basic Facts of Similarity

Similar figures have the same shape. The sizes may or may not be the same. Look at the similar figures. Now look at the matching angles. You can see that the angles are congruent.

Figures *A*, *B*, and *C* are similar figures. Figure *D* has a different shape than the other three. That means that figures *A*, *B*, *C*, and *D* are not similar. There is a way to help see that figures *A*, *B*, and *C* are similar. Think of turning figure *C* so that it is in the same position as figures *A* and *B*. By the way, figures *A* and *C* are also congruent.

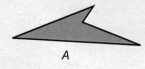 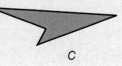 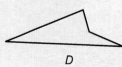
A B C D

Basic Facts of Similarity (cont.)

Look at the triangles at the top of the previous page. Some are different sizes. But only one has a different shape. Triangles *A*, *B*, *C*, and *D* are all similar. Triangle *E* does not belong.

Finding Missing Measurements

The two figures shown here are similar. We are going to see how to find the missing angle and side.

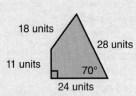
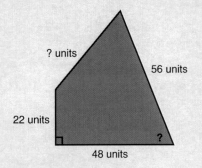

Angles: Matching angles of similar figures are congruent. So, the missing angle is 70°.

Sides: Each side of the first shape can be multiplied or divided by the same number to find each matching side of the second figure. For every side, we can multiply by 2 to find each matching side of the second figure. For example, 11 × 2 = 22. 28 × 2 = 56. And 24 × 2 = 48. So, the missing side is 18 × 2 = 36 units.

Similarity in Our Daily Lives

Look at the floor plan. This is a scale drawing of a house. Are these plans as large as a real house? Of course not! But all of the angles in the plans and the real thing are congruent. We can find the length of every segment in the drawing. Then we can multiply each segment by the same number to find the length of the real thing. Scale drawings are used to plan homes and cars. They can even be used to plan 100-story buildings!

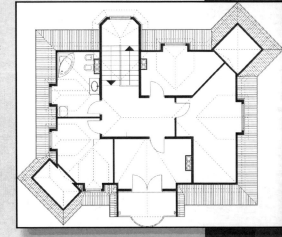

You Try It

The two triangles shown here are similar. Find the measure of the missing angle. Then find the length of the missing side. Now, draw a third triangle that is congruent with the first triangle.

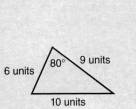
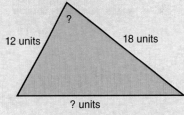

Congruent and Similar Figures

Which triangle does not belong?

Basic Facts of Congruence

Congruent segments have the same length.

Conceivably, you could lay one on top of the other and find that there is no difference in the two.

Congruent angles have the same measure.

Conceivably, you could lay one on top of the other and find that there is no difference in the two.

Two figures are congruent if they have the same size and shape. Conceivably, you could lay one on top of the other and find that there is no difference between them.

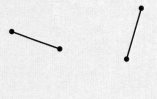

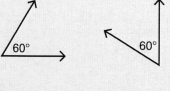

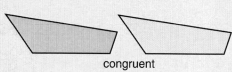
congruent

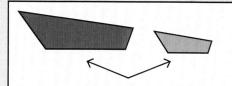
These figures are not congruent. They have different sizes.

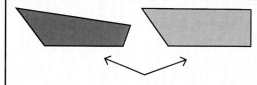
These figures are not congruent. They have different shapes.

Basic Facts of Similarity

Similar figures have the same shape, although the sizes may vary. As you look at the similar figures, notice that the matching angles of two similar figures are congruent.

Figures *A*, *B*, and *C* are similar figures. Figure *D* has a different shape than the other three. That means that figures *A*, *B*, *C*, and *D* are not similar. There is a way to help see that figures *A*, *B*, and *C* are similar. Think of turning figure *C* so that it is in the same position as figures *A* and *B*. By the way, figures *A* and *C* are also congruent.

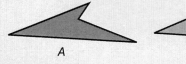 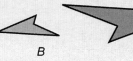 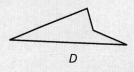

Basic Facts of Similarity (cont.)

Look at the triangles at the beginning of the previous page. While some may be smaller or larger than the others, only one triangle has a different shape altogether. Triangles *A*, *B*, *C*, and *D* are all similar to one another. Triangle E does not belong.

Finding Missing Measurements

The two figures shown here are similar. We are going to see how to find the missing angle and side.

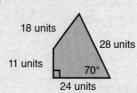
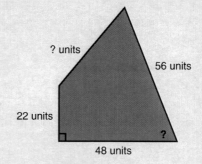

Angles: Matching angles of similar figures are congruent. So, the missing angle is 70°.

Sides: Every side of the first figure can be multiplied or divided by the same number in order to find each matching side of the second figure. For every side we do know, we can multiply by 2 to find each matching side of the second figure. For example, 11 × 2 = 22. 28 × 2 = 56. And 24 × 2 = 48. So, the missing side is 18 × 2 = 36 units.

Similarity in Our Daily Lives

Look at the scale drawing of the floor plan of a house. Are these plans as large as a real house? Of course not! But all of the angles in the plans and in the real thing are congruent. If you know the length of every segment in the drawing, you can multiply each segment by the same number to find the length of the real thing. Scale drawings are used in the planning of homes, cars, and even 100-story buildings!

You Try It

The two triangles shown here are similar. Find the measure of the missing angle and the length of the missing side. Then, draw a third triangle so the first and third triangles are congruent.

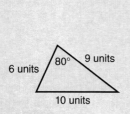
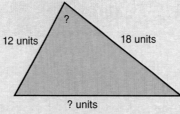

74

(E)

#50717—Leveled Texts for Mathematics: Geometry © Shell Education

Congruent and Similar Figures

Of the five triangles below, which one does not belong?

Basic Facts of Congruence

Congruent segments have the same length.

Conceivably, you could lay one on top of the other and find that there is no difference between the two.

Congruent angles have the same measure.

Conceivably, you could lay one on top of the other and find that there is no difference between the two.

Two figures are congruent if they have the same size and shape. Conceivably, you could lay one figure on top of the other and find that there is no difference between them.

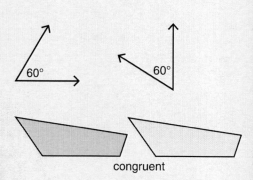

congruent

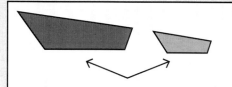

These figures are not congruent. They have different sizes.

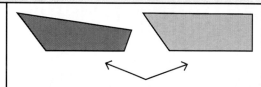

These figures are not congruent. They have different shapes.

Basic Facts of Similarity

Similar figures always have the same shape, although their sizes may or may not be comparable. As you examine the similar figures, notice that the matching angles of two similar figures are congruent.

Figures *A*, *B*, and *C* are similar figures. Figure *D* has a different shape than the other three. That means that figures *A*, *B*, *C*, and *D* are not similar. There is a way to help see that figures *A*, *B*, and *C* are similar. Think of turning figure *C* so that it is in the same position as figures *A* and *B*. By the way, figures *A* and *C* are also congruent.

Basic Facts of Similarity (cont.)

Let's reexamine the five triangles at the beginning of the previous page. Whereas they may each be a different size, only one triangle has a completely different shape. Triangles A, B, C, and D are all similar to one another, but you can see that triangle E does not belong.

Finding Missing Measurements

Because the two figures shown here are similar, we can easily find the measurement of the missing angle and side.

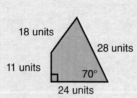
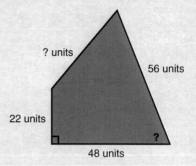

Angles: Since matching angles of similar figures are congruent, it follows that the missing angle is 70°.

Sides: Every side of the first figure can be multiplied or divided by the same number in order to determine each matching side of the second figure. For every side, we can multiply by 2 to find each matching side of the second figure. For example, $11 \times 2 = 22$. $28 \times 2 = 56$. And $24 \times 2 = 48$. So, the missing side is $18 \times 2 = 36$ units.

Similarity in Our Daily Lives

Obviously, a scale drawing of a house floor plan, like the one represented here, is not as large as the real house. However, all the angles in the plan and the real thing are congruent. Consequently, if you know the length of every segment in the drawing, you can multiply each segment by the same number to find the proportional length of the real thing. Scale drawings are used when planning homes, cars, and even 100-story buildings!

You Try It

These two triangles are similar. Find the measure of the missing angle and the length of the missing side. Then, draw a third triangle with this criteria: the first and third triangles must be congruent.

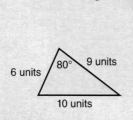
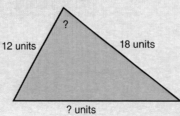

(E)

#50717—Leveled Texts for Mathematics: Geometry © Shell Education

Understanding 3-D Shapes

Think about a cereal box. How many rectangles does one box have?

Basic Facts

Imagine living in zero dimensions. You live your life on a point. You cannot move.

•

Imagine living in one dimension. You spend your life on a line or an angle. You can only move forward or backward. You cannot get around someone next to you.

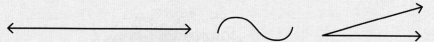

Think about living in two dimensions. Shapes like triangles and circles are now possible. Things can now have length and width. Things can have area. Think of a plane like a flat sheet of paper that goes on and on. You can now get around your neighbors. But you cannot move up or down.

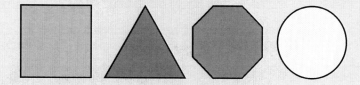

We live in three dimensions. We have length, width, and height. Our bodies have volume. We exist in the forms you see every day. We can move around objects. We can also move up and down. We can hold things because three-dimensional objects have thickness.

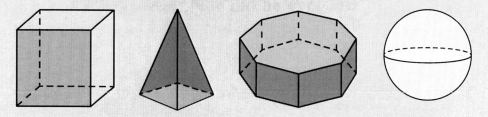

Did you know that the cereal box is a three-dimensional object with six sides?

Considering Solids

This three-dimensional shape has 6 faces. A **face** is the part of the shape that is flat.

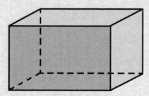

When two faces meet, a line segment is formed. Faces meet at the **edges**. A rectangular prism has 12 edges.

When edges meet, they meet at a point. Those points are the **vertices**. A single point where the edges meet is called a vertex. A rectangular prism has 8 vertices.

Counting Faces, Edges, and Vertices

The chart below shows solid shapes. As you look at each shape, make sure you can see why it has the number of faces, edges, and vertices that are listed.

Name	Rectangular Solid	Hexagonal Solid	Rectangular Pyramid	Triangular Pyramid	Sphere	Cylinder
Faces	6	8	5	4	0	2
Edges	12	18	8	6	0	0
Vertices	8	12	5	4	0	0

3-D Shapes in Our Daily Lives

We live in a three-dimensional world. Three-dimensional shapes are all around us. Did you know that the huge pyramids in Egypt were built over 4,500 years ago? The Great Pyramid of Giza was once 485 feet tall. It is now 450 feet tall due to erosion.

You Try It

Look at the solid below. How many faces, edges, and vertices does it have?

(6 rectangles)

Understanding 3-D Shapes

Think about a cereal box. How many rectangles does one box have?

Basic Facts

Imagine living in zero dimensions. You live your life on a point. You cannot move.

•

Imagine living in one dimension. You spend your life on a line or an angle. You can only move forward or backward. You cannot get around someone next to you.

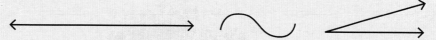

Think about living in two dimensions. Shapes like triangles and circles are now possible. Things can now have length and width. Things can have area. Your life can be on a plane. It would be like living on a flat sheet of paper that goes on and on. You can now get around your neighbors. But you cannot move up or down.

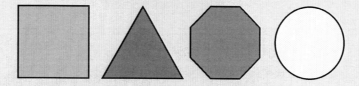

We live in three dimensions. We have length, width, and height. Our bodies have volume. We exist in the forms you see every day. We can move around objects. We can also move up and down. We can hold things because three-dimensional objects have thickness.

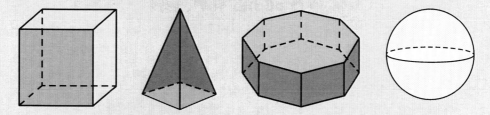

Did you realize that the cereal box is a three-dimensional object with six sides?

Considering Solids

This three-dimensional shape has 6 faces. A **face** is the part of the shape that is flat.

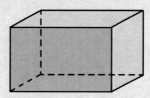

When two faces meet, a line segment is formed. Faces meet at the **edges**. A rectangular prism has 12 edges.

When edges meet, they meet at a point. Those points are the **vertices**. A single point where the edges meet is called a vertex. A rectangular prism has 8 vertices.

Counting Faces, Edges, and Vertices

The chart below shows solid shapes. As you look at each shape, make sure you can see why it has the number of faces, edges, and vertices that are listed.

Name	Rectangular Solid	Hexagonal Solid	Rectangular Pyramid	Triangular Pyramid	Sphere	Cylinder
Faces	6	8	5	4	0	2
Edges	12	18	8	6	0	0
Vertices	8	12	5	4	0	0

3-D Shapes in Our Daily Lives

We live in a three-dimensional world. Three-dimensional shapes are all around us. Did you know that the huge pyramids in Egypt were built over 4,500 years ago? The Great Pyramid of Giza was once 485 feet tall. It is now 450 feet tall due to erosion.

You Try It

Look at the solid below. How many faces, edges, and vertices does it have?

(6 rectangles)

Understanding 3-D Shapes

How many rectangles does it take to make a cereal box?

Basic Facts

Imagine living in zero dimensions. Your entire life is only on a point. You cannot go anywhere. No motion is possible.

•

Imagine living in one dimension. Your entire life might be on a line, a wavy line, or an angle. You can only move forward or backward. If people are next to you, you can never go around them.

Imagine living in two dimensions. Figures such as triangles and circles are now possible. Your body and everything around you can now have length and width. It is now possible for everything to have area. Your entire life can be on a two-dimensional plane. It would be like your entire life being on an endless, flat sheet of paper. You can now get around your neighbors, but you cannot go above or below anything.

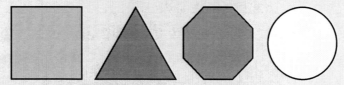

We live in three dimensions. We have length, width, and height. Our bodies have volume. We exist in the forms you see every day. We can move around objects. We can also move up and down. We can actually hold objects because three-dimensional objects have thickness.

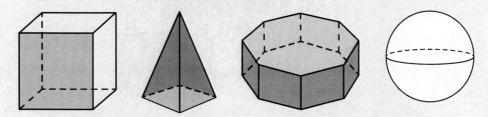

Did you realize that the cereal box is a three-dimensional object with six sides?

Considering Solids

This three-dimensional shape has 6 faces. A **face** is the part of the figure that is flat.

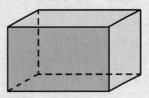

When two faces meet, a line segment is formed. Faces meet at the **edges**. A rectangular prism has 12 edges.

When edges meet, they meet at a point. Those points are called the **vertices**. A single point where the edges meet is called a vertex. Altogether, a rectangular prism has 8 vertices.

Counting Faces, Edges, and Vertices

Some solid figures are shown on the chart below. As you look at each figure, make sure you can see why it has the number of faces, edges, and vertices that are listed.

Name	Rectangular Solid	Hexagonal Solid	Rectangular Pyramid	Triangular Pyramid	Sphere	Cylinder
Faces	6	8	5	4	0	2
Edges	12	18	8	6	0	0
Vertices	8	12	5	4	0	0

3-D Shapes in Our Daily Lives

We live in a three-dimensional world, so three-dimensional shapes are all around us. Did you know that the huge pyramids in Egypt were built over 4,500 years ago? The Great Pyramid of Giza was once 485 feet tall, but is now 450 feet tall because of erosion.

You Try It

How many faces, edges, and vertices does the solid below have?

(6 rectangles)

Understanding 3-D Shapes

How many rectangles are needed to make a cereal box?

Basic Facts

Imagine living in zero dimensions. Your entire life would be lived only on a point. You could not go anywhere and motion would be impossible.

•

Imagine living in one dimension: your entire life would be lived on a line, a wavy line, or an angle. You could move forward or backward, but you could never go around someone who was alongside of you.

Imagine living in two dimensions, among figures such as triangles and circles. Your body and everything around you could now have length and width, and it would be possible for everything to have area. Your entire life could exist on a flat plane, similar to a sheet of paper that goes on forever. You could now get around your neighbors, but you could not go above or below anything.

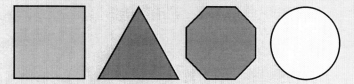

In reality, we live in three dimensions. We have length, width, and height, and our bodies have volume. We exist in the forms you see every day. We can move around objects, and we can also move up and down. We can actually hold objects because three-dimensional objects have thickness.

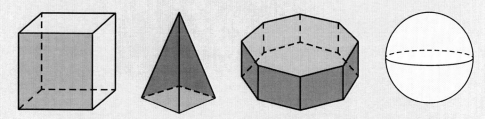

Did you recognize that the cereal box is a three-dimensional object with six sides?

Considering Solids

This three-dimensional shape has 6 faces. **Faces** are the parts of a figure that are flat.

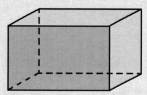

When two faces meet at the **edges**, a line segment is formed. A rectangular prism has 12 edges.

When edges meet, they meet at points called **vertices**. A single point where the edges meet is called a vertex. Altogether, a rectangular prism has 8 vertices.

Counting Faces, Edges, and Vertices

Some solid figures are shown on the chart. As you examine each figure, consider the rationale for why the particular number of faces, edges, and vertices are listed.

Name	Rectangular Solid	Hexagonal Solid	Rectangular Pyramid	Triangular Pyramid	Sphere	Cylinder
Faces	6	8	5	4	0	2
Edges	12	18	8	6	0	0
Vertices	8	12	5	4	0	0

3-D Shapes in Our Daily Lives

We live in a three-dimensional world, so three-dimensional shapes are all around us. Did you know that the huge pyramids in Egypt were built over 4,500 years ago? The Great Pyramid of Giza was once 485 feet tall, but is now 450 feet tall because of erosion.

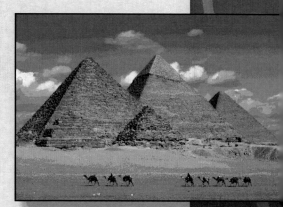

You Try It

How many faces, edges, and vertices does the solid below have?

(6 rectangles)

Understanding Prisms

Look at these shapes. How are they all alike?

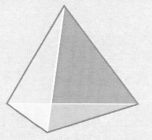 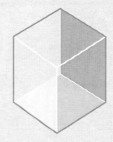 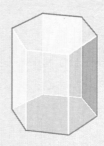 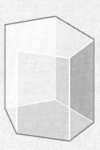

Basic Facts

A **polyhedron** is a shape. It is a three-dimensional shape. It is made up of faces. Every face is a polygon. It has no curves.

A **prism** is a polyhedron. It has two faces that are congruent and parallel to each other. The two faces are called the **bases** of the prism.

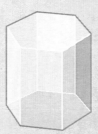 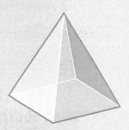

Prisms are named for their bases.

Triangular Prism	Rectangular Prism	Pentagonal Prism	Hexagonal Prism

Prisms and Nets

The prism shown is a pentagonal prism. It has 7 faces. It has 15 edges. It has 10 vertices.

Two faces are pentagons. They are the two parallel and congruent bases.

What if you could cut the prism along some of the edges? You could make the prism lie flat. You would end up with a **net** of the prism. What if a net were drawn on a sheet of paper? You could cut along the solid lines and fold along the dashed lines to form the shape. The net to the right is a net of a pentagonal prism.

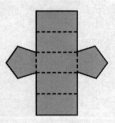

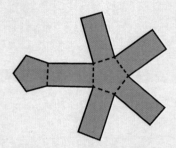

There is more than one possible net for any shape. The net to the left is also a net of a pentagonal prism. Notice that both nets are made up of 7 faces. You can see the 2 pentagons that make up each base. You can also see the 5 rectangles.

More Nets

The rectangular prism to the left has 6 faces, 12 edges, and 8 vertices. It is made up of 4 congruent rectangles and 2 squares. The two nets you see below are both nets that can form the box that is to the left.

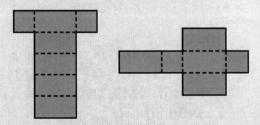

Prisms in Our Daily Lives

Did you know that the rainbow is made up of every color that we can see? You can make your own rainbow by shining a light through a glass, triangular prism. The prism bends the light so all the colors of the rainbow shine out.

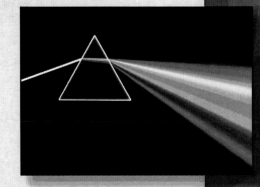

You Try It

Draw a net of a triangular prism.

Understanding Prisms

What do all of these shapes have in common?

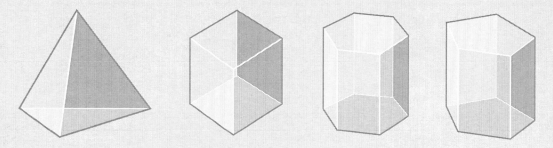

Basic Facts

A **polyhedron** is a three-dimensional figure. It is made up of faces. It has no curves and every face is a polygon.

A **prism** is a polyhedron. It has two faces that are congruent and parallel to one another. The two faces are called the **bases** of the prism.

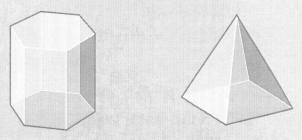

Prisms are named for their bases.

Triangular Prism	Rectangular Prism	Pentagonal Prism	Hexagonal Prism

Prisms and Nets

The prism shown is a pentagonal prism. It has 7 faces. It has 15 edges. It has 10 vertices.

Two faces are pentagons. They are the two parallel and congruent bases.

What if you could cut the prism along some of the edges? You could make the prism lie flat. You would end up with a **net** of the prism. What if a net were drawn on a sheet of paper? You could cut along the solid lines and fold along the dashed lines to form the shape. The net to the right is a net of a pentagonal prism.

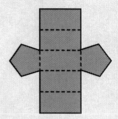

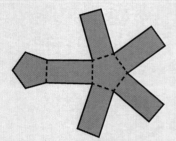

There is more than one possible net for any shape. The net to the left is also a net of a pentagonal prism. Notice that both nets are made up of 7 faces. You can see the two pentagons that make up each base. You can also see the 5 rectangles.

More Nets

The rectangular prism to the left has 6 faces, 12 edges, and 8 vertices. It is made up of 4 congruent rectangles and 2 squares. The two nets you see below are both nets that can form the box that is to the left.

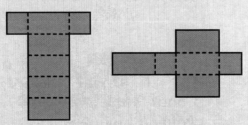

Prisms in Our Daily Lives

Did you know that the rainbow is made up of every color that we can see? You can make your own rainbow by shining a light through a glass, triangular prism. The prism bends the light so all the colors of the rainbow shine out.

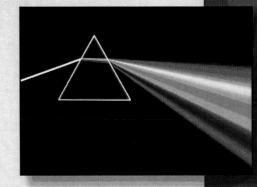

You Try It

Draw a net of a triangular prism.

Understanding Prisms

What is the commonality among all of these shapes?

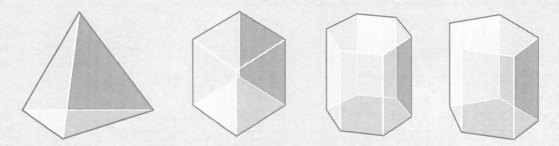

Basic Facts

A **polyhedron** is a three-dimensional figure that is made up entirely of faces. It has no curves, and every face is a polygon.

A **prism** is a polyhedron that has two faces that are congruent and parallel to one another. The two faces are called the **bases** of the polyhedron, as in a prism.

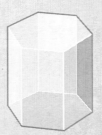 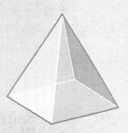

Prisms are named according to the shape of their bases.

Triangular Prism	Rectangular Prism	Pentagonal Prism	Hexagonal Prism

Prisms and Nets

The prism shown is a pentagonal prism. It has 7 faces, 15 edges, and 10 vertices.

The two faces that are pentagons are the two parallel and congruent bases.

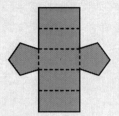

Imagine cutting the prism along some of the edges so that the prism is able to lie flat. The result is a **net** of the prism. If a net were drawn on a sheet of paper, then you could cut along the solid lines and fold along the dashed lines to form the figure. The net to the right is a net of a pentagonal prism.

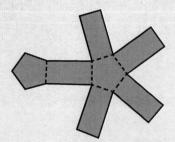

There is more than one possible net for any figure. The net to the left is also a net of a pentagonal prism. Notice that both nets are made up of 7 faces. You can see the 2 pentagons that make up each base as well as the 5 rectangles.

More Nets

The rectangular prism to the left has 6 faces, 12 edges, and 8 vertices. It is made up of 4 congruent rectangles and 2 squares. The two nets you see below are both nets that can form the box shown to the left.

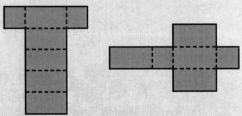

Prisms in Our Daily Lives

Did you know that the rainbow is actually made up of every color that we can see? You can make your own rainbow by shining a light through a glass, triangular prism. The prism bends the light so that all the colors of the rainbow shine out.

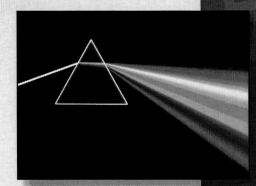

You Try It

Draw a net of a triangular prism.

Understanding Prisms

Examine these shapes and determine what they all have in common.

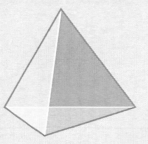 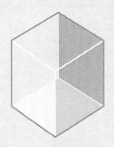 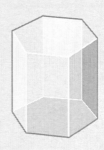 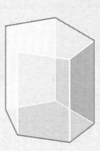

Basic Facts

A **polyhedron** is a three-dimensional figure that consists entirely of faces, all of which are polygons.

A **prism** is a polyhedron that has two faces. These faces are called the **bases** of the prism and they are congruent and parallel.

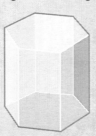 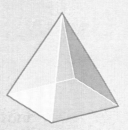

Prisms are named for their bases.

Triangular Prism	Rectangular Prism	Pentagonal Prism	Hexagonal Prism

Prisms and Nets

This is a pentagonal prism with 7 faces, 15 edges, and 10 vertices.

The two faces that are pentagons are the two parallel and congruent bases.

Imagine cutting the prism along some of the edges so that the prism is able to lie flat. The result is a **net** of the prism. If a net were drawn on a sheet of paper, you would be able to cut along the solid lines and fold along the dashed lines to form the figure. The net to the right is a net of a pentagonal prism.

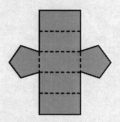

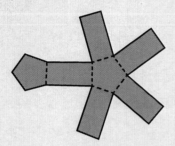

There is more than one possible net for any figure. The net to the left is also a net of a pentagonal prism. You may notice that both nets consist of 7 faces, so you can see the 2 pentagons that make up each base as well as the 5 rectangles.

More Nets

The rectangular prism to the left has 6 faces, 12 edges, and 8 vertices. It is made up of 4 congruent rectangles and 2 squares. Did you notice that both nets shown below could be assembled to form the box shown to the left?

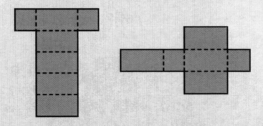

Prisms in Our Daily Lives

Have you considered that the rainbow actually consists of every color that we have the ability to see? You can manufacture your own rainbow by shining a light through a glass, triangular prism, which bends the light, allowing every color of the rainbow to radiate out.

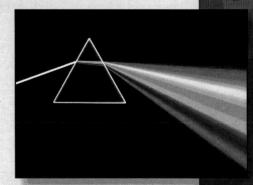

You Try It

Draw a net of a triangular prism.

#50717—Leveled Texts for Mathematics: Geometry © Shell Education

The Coordinate Plane

Look at the number line below. How far apart are the two points?

Basic Facts

We can draw two number lines so that they are perpendicular to each other. When we do that, we form a **coordinate plane**.

The two number lines in the coordinate plane are called the **axes**. One number line forms the horizontal axis (↔). This is called the ***x*-axis**. The other number line forms the vertical axis (↕). This is the ***y*-axis**.

Look at the point shown on the right. It is at the coordinates (2, 3). The numbers tell us where the point may be found in the coordinate plane. Follow the dashed lines. You can see that the first coordinate is at 2. The second one is at 3.

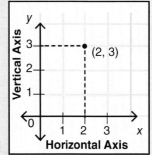

Horizontal Coordinate: The first number in an ordered pair tells how far to the right or left the point is from the vertical axis.

Vertical Coordinate: The second number in an ordered pair tells how far up or down the point is from the horizontal axis.

We write the coordinates of a point in an **ordered pair**. We write it as follows: (horizontal coordinate, vertical coordinate). This can also be thought of as (*x* value, *y* value).

Plotting Points and Moving Among Points

We can see movement from one point to the next. Let's start at point *A*. We can move four units to the right. That gets us to point *B*. From point *B*, we can move two units up. That gets us to point *C*. From point *C*, we can move three units to the right. That gets us to point *D*.

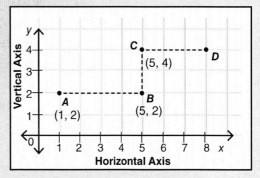

Look at point *D*. Let's find its coordinates. We know that ordered pairs are written as (*x*, *y*) or (horizontal coordinate, vertical coordinate).

First, trace down to the *x*-axis from point *D*. You can see that the horizontal coordinate is 8. Then, trace across to the *y*-axis from *D*. You can see that the vertical coordinate is 4. Point *D* is at (8, 4).

Distances in the Coordinate Plane

We can find distances between two points on a coordinate plane.

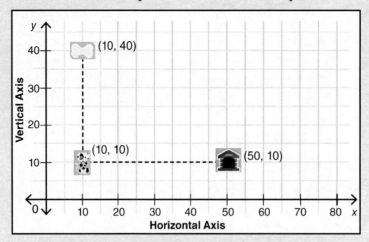

The dog wants its bone. It has to walk 30 units up to get it.

- Count by fives along the dashed line to find the distance of 30 units.

- From the dog at (10, 10) and the bone at (10, 40), we subtract the two *y* values (40 – 10) to find 30 units.

The Coordinate Plane in Our Daily Lives

"A picture is worth a thousand words." You may know that saying. The same is true for graphs. A graph helps us to understand the data that is shown.

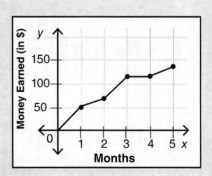

Graphs are used to help us see our height and weight over time. Businesses use graphs to see how much money was made each month. They can compare that with their costs. A graph that shows where a car is over time can show if the car is speeding up or slowing down!

You Try It

Look at the coordinate plane above. What are the coordinates of the doghouse? How far is the dog from the doghouse?

The Coordinate Plane

Look at the number line below. What is the distance between the two points?

Basic Facts

We can draw two number lines so that they are perpendicular to each other. When we do that, we form a **coordinate plane**.

The two number lines in the coordinate plane are called the **axes**. One number line forms the horizontal axis (↔). This is called the *x*-axis. The other number line forms the vertical axis (↕). This is the *y*-axis.

Look at the point shown on the right. It is at the coordinates (2, 3). The numbers tell us where the point may be found in the coordinate plane. Follow the dashed lines. You can see that the first coordinate is at 2. The second one is at 3.

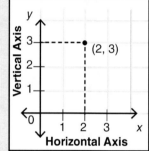

Horizontal Coordinate: The first number in an ordered pair tells how far to the right or left the point is from the vertical axis.

Vertical Coordinate: The second number in an ordered pair tells how far up or down the point is from the horizontal axis.

We write the coordinates of a point in an **ordered pair**. We write it as follows: (horizontal coordinate, vertical coordinate). This can also be thought of as (*x* value, *y* value).

Plotting Points and Moving Among Points

We can see movement from one point to the next. Let's start at point *A*. We can move four units to the right. That gets us to point *B*. From point *B*, we can move two units up. That gets us to point *C*. From point *C*, we can move three units to the right. That gets us to point *D*.

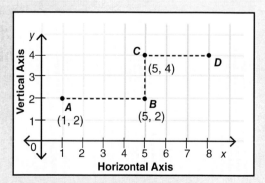

Look at point *D*. Let's find its coordinates. We know that ordered pairs are written as (*x*, *y*) or (horizontal coordinate, vertical coordinate).

First, trace down to the *x*-axis from point *D*. You can see that the horizontal coordinate is 8. Then, trace across to the *y*-axis from *D*. You can see that the vertical coordinate is 4. Point *D* is at (8, 4).

Distances in the Coordinate Plane

We can find distances between two points on a coordinate plane.

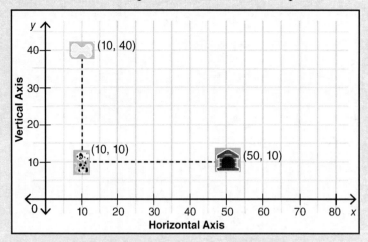

The dog wants its bone. It has to walk 30 units up to get it.

- Count by fives along the dashed line to find the distance of 30 units.

- From the dog at (10, 10) and the bone at (10, 40), we subtract the two *y* values (40 – 10) to find 30 units.

The Coordinate Plane in Our Daily Lives

"A picture is worth a thousand words." You may know that saying. The same is true for graphs. A graph helps us to understand the data that is shown.

Graphs are used to help us see our height and weight over time. Businesses use graphs to see how much money was made each month. They can compare that with their costs. A graph that shows where a car is over time can show if the car is speeding up or slowing down!

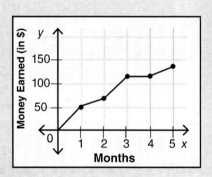

You Try It

Look at the coordinate plane above. What are the coordinates of the doghouse? How far is the dog from the doghouse?

The Coordinate Plane

Look at the number line below. What is the distance between the two points?

Basic Facts

If we draw two number lines so that they are perpendicular to each other, we form a **coordinate plane**.

The two number lines in the coordinate plane are called the **axes**. One number line forms the horizontal axis (↔). This is called the ***x*-axis**. The other number line forms the vertical axis (↕). This is the ***y*-axis**.

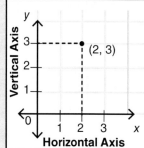

The point shown at the right is at the coordinates (2, 3). The numbers tell us where the point may be found in the coordinate plane. Follow the dashed lines to see that the first coordinate is at 2 and the second coordinate is at 3.

Horizontal Coordinate: The first number in an ordered pair tells how far to the right or left the point is from the vertical axis.

Vertical Coordinate: The second number in an ordered pair tells how far up or down the point is from the horizontal axis.

We write the coordinates of a point in an **ordered pair** as follows: (horizontal coordinate, vertical coordinate). This can also be thought of as (*x* value, *y* value).

Plotting Points and Moving Among Points

We can describe movement from one point to another. If we start at point *A*, we can move four units to the right to arrive at point *B*. From point *B*, we can move two units up to arrive at point *C*. From point *C*, we can move three units to the right to arrive at point *D*.

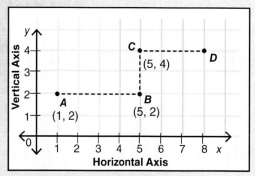

Let's find the coordinates of point *D*. Remember that ordered pairs are written as (*x*, *y*) or (horizontal coordinate, vertical coordinate).

If you trace down to the *x*-axis from point *D*, you can see that the horizontal coordinate is 8. If you trace across to the *y*-axis from *D*, you see that the vertical coordinate is 4. Point *D* is at (8, 4).

Distances in the Coordinate Plane

We can find distances between two points on a coordinate plane.

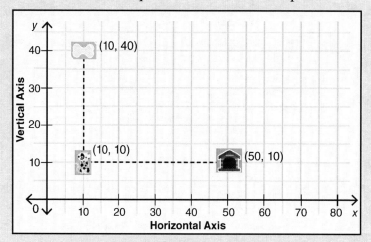

The dog has to walk 30 units up to get to the bone.

- Count by fives along the dashed line to find the distance of 30 units.

- From the dog at (10, 10) and the bone at (10, 40), we subtract the two *y* values (40 – 10) to find 30 units.

The Coordinate Plane in Our Daily Lives

You may have heard the saying, "A picture is worth a thousand words." The same is true for graphs. Looking at a graph helps us to understand the information being shown.

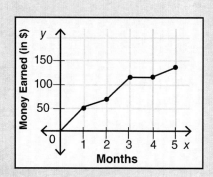

Graphs are used to help us see our height and weight over time. They help businesses compare the amount of money made each month with the cost. A graph of the position of a car over time can even show whether the car is speeding up or slowing down!

You Try It

Look at the coordinate plane above. What are the coordinates of the doghouse? What is the distance between the dog and the doghouse?

The Coordinate Plane

Look at the number line below and see if you can determine the distance between the two points.

Basic Facts

If we draw two number lines so that they are perpendicular to each other, we create a **coordinate plane**.

The two number lines in the coordinate plane are called the **axes**. The number line that forms the horizontal axis (↔) is called the ***x*-axis**. The number line that forms the vertical axis (↕) is called the ***y*-axis**.

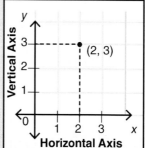

The point indicated at the right is at the coordinates (2, 3). The numbers tell us where the point may be found in the coordinate plane. Follow the dashed lines to see that the first coordinate is positioned at 2 and the second coordinate is at 3.

Horizontal Coordinate: The first number in an ordered pair tells how far to the right or left the point is from the vertical axis.

Vertical Coordinate: The second number in an ordered pair tells how far up or down the point is from the horizontal axis.

The coordinates of a point in an **ordered pair** are written as follows: (horizontal coordinate, vertical coordinate). This can also be thought of as (*x* value, *y* value).

Plotting Points and Moving Among Points

Coordinates help us describe movement from one point to another. If we start at point *A*, we can move four units to the right to arrive at point *B*. From point *B*, we can move two units up to arrive at point *C*, and from point *C*, moving three units to the right will get us to point *D*.

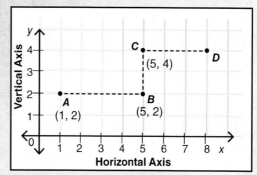

Now we're ready to determine the coordinates of point *D*. Remember that ordered pairs are written as (*x*, *y*) or (horizontal coordinate, vertical coordinate).

If you trace down to the *x*-axis from point *D*, you see that the horizontal coordinate is 8. If you trace across to the *y*-axis from *D*, you see that the vertical coordinate is 4. Point *D* is at (8, 4).

Distances in the Coordinate Plane

We can determine distances between two points on a coordinate plane.

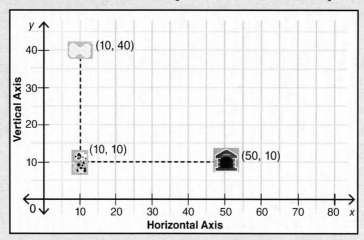

The dog has to walk 30 units up to get to the bone.

- Count by fives along the dashed line to find the distance of 30 units.

- From the dog at (10, 10) and the bone at (10, 40), we subtract the two *y* values (40 – 10) to find 30 units.

The Coordinate Plane in Our Daily Lives

You may have heard the saying, "A picture is worth a thousand words." The same is true for graphs. Studying a graph helps us to gain a better understanding of the information being shown.

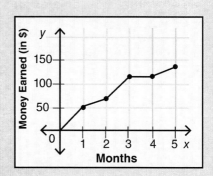

Graphs are used to help us see our height and weight over time. They help businesses compare the amount of money made each month with their expenses. A graph of the position of a car when compared with time can even show whether the car is speeding up or slowing down!

You Try It

Look at the coordinate plane above to determine the coordinates of the doghouse. What is the distance between the dog and the doghouse?

Circles

Who came up with the idea for the Round Table?

Basic Facts

A **circle** is a figure that lies in a plane. It is two dimensional. A circle is a set of points that are all the same distance from one point in the middle of the circle. That point is the **center**. A **radius** is a line segment. It can be drawn from the center to any point on the circle. A **diameter** is a line segment, too. It is drawn from one point on the circle to another. It passes through the center of the circle.

This is a circle. The points inside are not part of the circle. The points outside are not part of the circle. Only the points that are the same distance from the center are part of the circle.

The point in the middle is the center.

All points on the circle are the same distance from the center. Points A, B, and C are all 2 units from the center. Points D and E are, too.

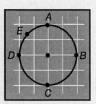

The radius is a line segment. It can be drawn from the center to the circle.

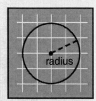

The diameter is a line segment. It is drawn from one point on the circle to another point on the circle and passes through the center.

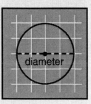

The **circumference** is how far it is around the circle. Think of the circle as a string. Stretch the string out. The length of the string is the circumference of the circle. We can measure the space inside the circle, too. That is called the **area**.

Understanding the Relationship

These two circles are the same size. The radius is shown on the first circle. The diameter is shown on the second circle. See that the radius is half the diameter. If we know one value, we can find the other.

radius = 2 units

diameter = 4 units

If we know the radius, we can find the diameter.

The radius above is 30 units. (r = 30)

Diameter = 2 × radius. (d = 2 × 30 = 60)

The diameter is 60 units.

If we know the diameter, we can find the radius.

The diameter above is 16 units. (d = 16)

Radius = diameter ÷ 2. (r = 16 ÷ 2 = 8)

The radius is 8 units.

Circles in Our Daily Lives

When we know about circles, it helps us learn about other things. Gears are a good example. If you move one gear, another gear can move faster or slower. It can even move in a different direction. Gears are used in clocks and watches. They are used in bikes, too!

You Try It

Draw a circle. Mark the center. Draw the radius. Mark it as 17 units. Now find the diameter.

Circles

Who came up with the idea for the Round Table?

Basic Facts

A **circle** is a figure that lies in a plane. It is two dimensional. A circle is a set of points that are all the same distance from one point in the middle of the circle. That point is the **center**. A **radius** is a line segment. It can be drawn from the center to any point on the circle. A **diameter** is a line segment, too. It is drawn from one point on the circle to another. It passes through the center of the circle.

This is a circle. The points inside are not part of the circle. The points outside are not part of the circle. Only the points that are the same distance from the center are part of the circle.

The point in the middle is the center.

All points on the circle are the same distance from the center. Points A, B, and C are all 2 units from the center. Points D and E are, too.

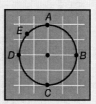

The radius is a line segment. It can be drawn from the center to the circle.

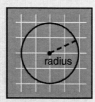

The diameter is a line segment. It is drawn from one point on the circle to another point on the circle and passes through the center.

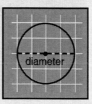

The **circumference** is how far it is around the circle. Think of the circle as a string. Stretch the string out. The length of the string is the circumference of the circle. We can measure the space inside the circle, too. That is called the **area**.

Understanding the Relationship

The two circles below are the same size. The radius is shown on the first circle. The diameter is shown on the second circle. See that the radius is half the diameter. If we know one value, we can find the other.

radius = 2 units

diameter = 4 units

If we know the radius, we can find the diameter.	If we know the diameter, we can find the radius.
The radius above is 30 units. ($r = 30$)	The diameter above is 16 units. ($d = 16$)
Diameter = 2 × radius. ($d = 2 \times 30 = 60$)	Radius = diameter ÷ 2. ($r = 16 \div 2 = 8$)
The diameter is 60 units.	The radius is 8 units.

Circles in Our Daily Lives

When we know about circles, it helps us learn about other things. Gears are a good example. If you move one gear, another gear can move faster or slower. It can even move in a different direction. Gears are used in clocks and watches. They are used in bikes, too!

You Try It

Draw a circle. Mark the center. Draw the radius. Mark it as 17 units. Now find the diameter.

(Sir Cumference)

Circles

Who invented the Round Table?

Basic Facts

A **circle** is a figure that lies in a plane. It is a two-dimensional figure. A circle is the set of points that are all the same distance from one point that lies inside the circle. That one point inside the circle is the **center**. A **radius** is any line segment that can be drawn from the center to any point on the circle. A **diameter** is any line segment that passes through the center as it is drawn from one point on the circle to another.

This is a circle. The points inside the figure are not part of the circle. The points outside the figure are not part of the circle. Only the points that are the same distance from the center are part of the circle.

The point in the middle is called the center.

All points on the circle are the same distance from the center. For example, points A, B, C, D, and E are all 2 units from the center.

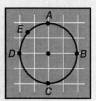

The radius of a circle is the line segment that can be drawn from the center to the circle.

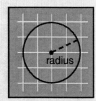

The diameter of a circle is the line segment that can be drawn through the center. It is drawn from one point on the circle to another point on the circle.

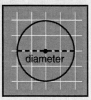

The **circumference** is the distance around the circle. Imagine the circle as a string. Stretch the string out so that it forms a line segment. The length of the string is the circumference of the circle. The **area** is the measurement of the space inside the circle.

Understanding the Relationship

The two circles shown below are the same size. The radius is given for one circle, and the diameter is given for the second. Notice that the radius is half the diameter. If we know one value, we can find the other.

radius = 2 units

diameter = 4 units

If we know the radius, we can find the diameter.

The radius above is 30 units. ($r = 30$)

Diameter = 2 × radius. ($d = 2 \times 30 = 60$)

The diameter is 60 units.

If we know the diameter, we can find the radius.

The diameter above is 16 units. ($d = 16$)

Radius = diameter ÷ 2. ($r = 16 \div 2 = 8$)

The radius is 8 units.

Circles in Our Daily Lives

Understanding circles is the key to understanding gears. If you move one gear, another gear can move faster, slower, or even in a different direction. Gears are used in clocks, watches, and even bicycles!

You Try It

Draw a circle. Mark the center. Draw the radius. Label the radius as 17 units. Find the diameter of the circle.

(Sir Cumference)

Circles

Who invented the Round Table?

Basic Facts

A **circle** is a two-dimensional figure that lies in a plane. A circle can be defined as the set of points that are all the same distance from one point that lies inside the middle of the circle, which is called the **center**. A **radius** is any line segment that can be drawn from the center of the circle to any point on the circle. A **diameter** is any line segment that passes across the center of the circle as it is drawn from one point on the circle to another point.

If you study this circle, you will notice that neither the points inside the figure nor the points outside the figure are part of the circle. Only the points that are the same distance from the center are part of the circle.

The point in the middle is called the center.

All points on the circle are the same distance from the center. For example, points A, B, C, D, and E are all 2 units from the center.

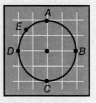

The radius of a circle is the line segment that can be drawn from the center to the circle.

The diameter of a circle is the line segment that can be drawn from one point on the circle, across the center, to another point on the circle.

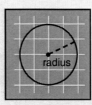

The **circumference** is defined as the distance around the circle. Imagine the circle as a string that is stretched out so that it forms a line segment. The length of the string is the circumference of the circle. The measurement of the space inside the circle is called the **area**.

Understanding the Relationship

The two circles shown below are identical in size. The radius is shown in the first circle, and the diameter is shown in the second circle. Notice that the radius is half the diameter. If we know one value, we can find the other.

radius = 2 units	diameter = 4 units
If we know the radius, we can find the diameter. The radius above is 30 units. ($r = 30$) Diameter = 2 × radius. ($d = 2 \times 30 = 60$) The diameter is 60 units.	If we know the diameter, we can find the radius. The diameter above is 16 units. ($d = 16$) Radius = diameter ÷ 2. ($r = 16 \div 2 = 8$) The radius is 8 units.

Circles in Our Daily Lives

Understanding circles is critical to understanding gears. If you move one gear, another gear can move faster, slower, or even in a different direction. Gears are used in clocks, watches, machinery, and even bicycles!

You Try It

Draw a circle and identify its center. Next, draw the radius and label it as 17 units. Finally, use that information to determine the diameter of the circle.

(Sir Cumference)

Symmetry

Look at this butterfly. Is it symmetrical? How do you know?

Basic Facts

Here is a good way to know if a figure has **symmetry**. Can you fold the figure in half so that the two sides are an exact match? If the answer is yes, then the figure has symmetry. The type of symmetry seen with folding is called **line symmetry**. Do you see the dashed line through the heart on the left? That is the **line of symmetry**. What if you folded along the line? The two sides of the heart would be an exact match! Look at the heart on the right. It is not symmetrical. What if you folded it along the dashed line? The two sides would not match.

Identifying Lines of Symmetry

Look at each figure or drawing below. Each has at least one line of symmetry. Notice that some lines are vertical. Some lines are horizontal. Some lines are even diagonal. As you look at them, think of folding along the line of symmetry. See how the two sides exactly match.

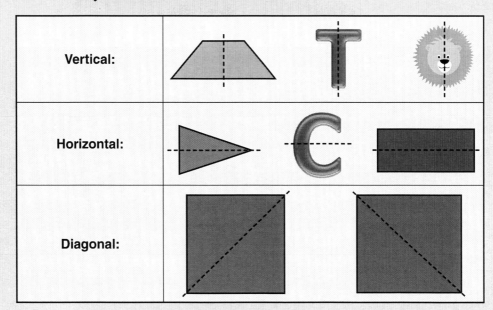

More Than One Line of Symmetry

Sometimes figures can have more than one line of symmetry.

A rectangle has two lines of symmetry. One line is vertical. The other line is horizontal.

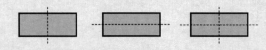

A square has a vertical line of symmetry. It also has a horizontal line of symmetry. It has two diagonal lines of symmetry. So, a square has four lines of symmetry.

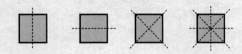

A regular pentagon has five lines of symmetry. A regular hexagon has six.

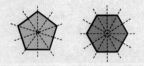

A circle has an endless number of lines of symmetry. Only a few of the lines are shown here. Notice how they all pass though the center of the circle.

The key to a line of symmetry is matching. If you can fold a figure in half so that there is no difference in the two sides, then it has a line of symmetry. If there is any difference at all, then there is no line of symmetry. Did the two sides of your butterfly match? If so, then your butterfly is symmetrical. If not, then your butterfly is not symmetrical.

Symmetry in Our Daily Lives

Symmetry is all around us. It is in nature. It is in art. It is even in buildings. Look at the Taj Mahal. It is in India. We tend to like things that are symmetrical. This is why people use symmetry in design and art.

You Try It

Draw a regular octagon. How many lines of symmetry does it have?

(yes)

Symmetry

Look at this butterfly. Is it symmetrical? How do you know?

Basic Facts

Here is a good way to know if a figure has **symmetry**. Can you fold the figure in half so that the two sides are an exact match? If the answer is yes, then the figure has symmetry. The type of symmetry seen with folding is called **line symmetry**. Do you see the dashed line through the heart on the left? That is the **line of symmetry**. What if you folded along the line? The two sides of the heart would be an exact match! Look at the heart on the right. It is not symmetrical. What if you folded along the dashed line? The two sides would not match.

Identifying Lines of Symmetry

Look at each figure or drawing below. Each figure has at least one line of symmetry. Notice that some lines are vertical. Some lines are horizontal. Some lines are even diagonal. As you look at them, think of folding along the line of symmetry. See how the two sides exactly match.

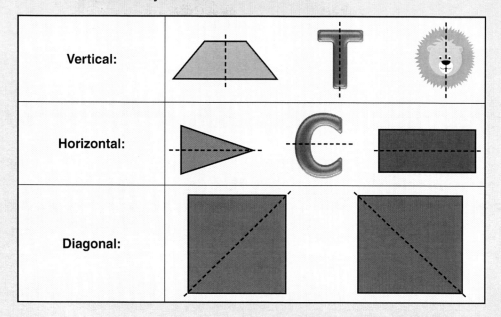

More Than One Line of Symmetry

Sometimes figures can have more than one line of symmetry.

A rectangle has two lines of symmetry. One line is vertical. The other line is horizontal.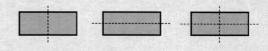

A square has a vertical line of symmetry. It also has a horizontal line of symmetry. It has two diagonal lines of symmetry. So, a square has four lines of symmetry.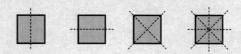

A regular pentagon has five lines of symmetry. A regular hexagon has six.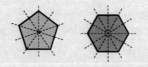

A circle has an endless number of lines of symmetry. Only a few of the lines are shown here. Notice how they all pass though the center of the circle.

The key to a line of symmetry is matching. If you can fold a figure in half so that there is no difference in the two sides, then it has a line of symmetry. If there is any difference at all, then there is no line of symmetry. Did the two sides of your butterfly match? If so, then your butterfly is symmetrical. If not, then your butterfly is not symmetrical.

Symmetry in Our Daily Lives

Symmetry is all around us. It is in nature. It is in art. It is even in buildings. Look at the Taj Mahal. It is in India. We tend to like things that are symmetrical. This is why people use symmetry in design and art.

You Try It

Draw a regular octagon. How many lines of symmetry does it have?

(yes)

#50717—Leveled Texts for Mathematics: Geometry © Shell Education

Symmetry

Is this butterfly symmetrical? How do you know?

Basic Facts

Figures have **symmetry** if it is possible to fold the figure in half so that the two sides exactly match. The particular type of symmetry seen with folding is called **line symmetry**. The dashed line passing through the heart on the left is the **line of symmetry**. If you fold along the line, the two sides of the heart match exactly. The heart on the right is not symmetrical. If you try to fold along the dashed line, the two sides do not exactly match.

Identifying Lines of Symmetry

Look at each figure or drawing below. Each figure has at least one line of symmetry. Notice that some lines are vertical. Some lines are horizontal. Some lines are even diagonal. As you look at each one, think of folding along the line of symmetry. See how the two sides exactly match.

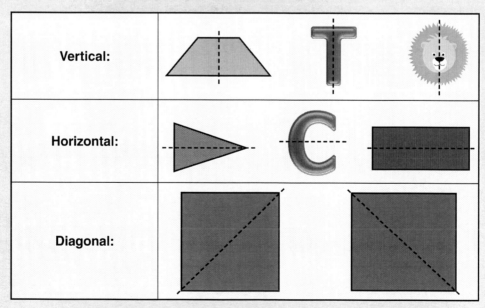

More Than One Line of Symmetry

Sometimes figures can have more than one line of symmetry.

A rectangle has two lines of symmetry. One line is vertical and the other line is horizontal.

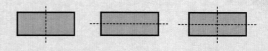

A square has a vertical line of symmetry. It also has a horizontal line of symmetry. It even has two diagonal lines of symmetry. So, a square has four lines of symmetry.

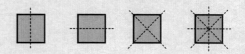

A regular pentagon has five lines of symmetry. A regular hexagon has six.

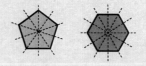

A circle has an endless number of lines of symmetry. Only a few of the lines are shown here. Notice that they all pass though the center of the circle.

The key to a line of symmetry is matching. If you can fold a figure in half so that there is no difference in the two sides, then it has a line of symmetry. If there is any difference at all, then the line is not a line of symmetry. Did the two sides of your butterfly match? If so, then your butterfly is symmetrical. If not, then your butterfly is not symmetrical.

Symmetry in Our Daily Lives

Symmetry is found throughout the world around us. It is in nature, artwork, and even the buildings we see every day. It is in structures such as the Taj Mahal in India. Without even thinking of it, we tend to like images that are symmetrical. Designers, architects, and artists use symmetry to create images that are pleasing to us.

You Try It

Draw a regular octagon. How many lines of symmetry does it have?

(yes)

Symmetry

Is this butterfly symmetrical? How do you know?

Basic Facts

Figures have **symmetry** if it is possible to fold the figure in half so that the two sides exactly match. The particular type of symmetry seen with folding is called **line symmetry**. The dashed line passing through the heart on the left is the **line of symmetry**. If you fold along the line, the two sides of the heart match exactly. The heart on the right is not symmetrical because if you try to fold along the dashed line, the two sides do not exactly match.

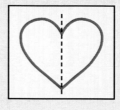 

Identifying Lines of Symmetry

Examine each figure or drawing below, each with one of its lines of symmetry. Notice that some lines are vertical, while others are horizontal and even diagonal. As you study each one, imagine folding along the line of symmetry and see how the two sides exactly match.

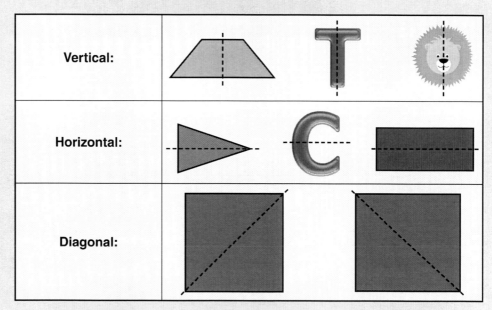

More Than One Line of Symmetry

Sometimes figures can have multiple lines of symmetry.

A rectangle has two lines of symmetry—one vertical and the other horizontal.

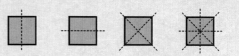

A square has four lines of symmetry: a vertical line, a horizontal line, and two diagonal lines.

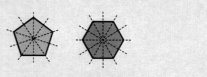

A regular pentagon has five lines of symmetry, while a regular hexagon has six.

A circle has an infinite number of lines of symmetry. Only a few of the lines are shown here. Notice that they all pass though the center of the circle.

The key to a line of symmetry is matching. If you can fold a figure so that there is no difference in the two sides, then it has a line of symmetry. If there is any difference at all, then the line is not a line of symmetry. Did the two sides of your butterfly match? If so, then your butterfly is symmetrical. However, if they did not match, then your butterfly is not symmetrical.

Symmetry in Our Daily Lives

Symmetry is found in the world around us: in nature, in artwork, and even in the buildings we see every day. You can see it in structures such as the Taj Mahal in India. You might not even realize it, but you probably like images that are symmetrical. Designers, architects, and artists understand this, which is why they often use symmetry to create images that are pleasing to us.

You Try It

Draw a regular octagon. How many lines of symmetry does it have?

(yes)

Reflections

Have you ever looked at words in a mirror? Are the words easy to read? Do they seem backwards?

Basic Facts

A reflection is a type of transformation. A **transformation** is an adjustment of a figure. The figure can be the same size and shape. It may be in a different position. The figure can be stretched. It can be smaller.

A **reflection** is when a figure is flipped. There is a line to flip the figure across. This is called the **line of reflection**. The figure and its reflection are the same size and shape. That means that they are congruent. Matching points are the same distance from the line of reflection. Look at the figures below. Each point *A* is five units from the line of reflection. Each point *B* is one unit from the line of reflection. Each point *C* is three units from the line of reflection.

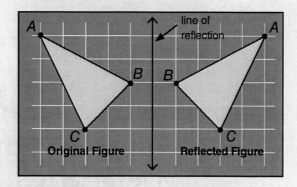

Look at the two triangles above. Do you see the symmetry in each picture? The triangles have **line symmetry**. There is another name for line symmetry. It is **reflection symmetry**.

Sample Reflections

Look at the figures below. In each one, the image on one side of the line is the original. The image on the other side of the line is the reflection. Notice the congruence of the two sides. See that the second image is a "flipped" version of the original.

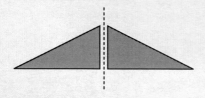 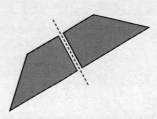 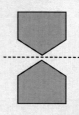

Reflecting an Image on a Grid

Look at the left triangle on the grid below. How do we find its reflection?

First, draw a line of reflection. In this case, the line of reflection is already drawn.

Next, choose a point. Find the distance between that point and the line of reflection. Point *A* is five units to the left of the line of reflection. That means that the reflected point will be five units to the *right* of the line. Plot point *A'* five units to the right of the line of reflection. Make sure it is at the same height.

Point *B* is three units to the left of the line. That means that the reflected point will be three units to the *right* of the line. Plot point *B'* three units to the right of the line of reflection.

Point *C* is one unit to the left of the line. That means that the reflected point will be one unit to the *right* of the line. Plot point *C'* one unit to the right of the line of reflection. Connect all the points.

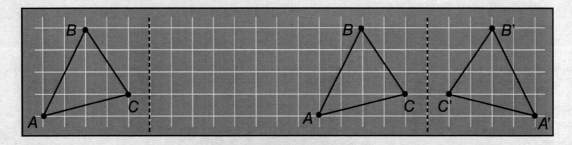

Reflections in Our Daily Lives

At the start of the lesson you thought about what words look like in a mirror. Each is a reflection of the original. Have you ever noticed the front of an ambulance? The word *ambulance* is a reflection. It is flipped. This way, a driver in front of an oncoming ambulance can read the word in the rearview mirror.

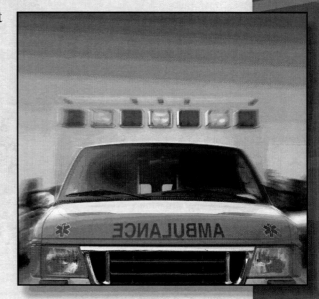

You Try It

Draw a reflection of the figure below.

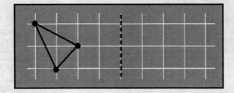

Reflections

Have you ever looked at words in a mirror? Are those words easy to read? Or do they seem backwards?

Basic Facts

A reflection is a type of transformation. A **transformation** is an adjustment of a figure. The figure can be the same size and shape. But it may be in a different position. The figure can be stretched or shrunk.

A **reflection** is when a figure is flipped. There is a line that the figure is flipped across. This is called the **line of reflection**. The figure and its reflection are the same size and shape. That means that they are congruent. Matching points are the same distance from the line of reflection. Look at the figures below. Each point A is five units from the line of reflection. Each point B is one unit from the line of reflection. Each point C is three units from the line of reflection.

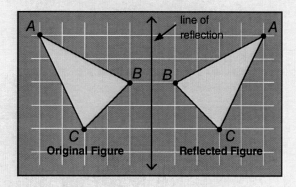

Look at the two triangles above. Do you see the symmetry in each picture? The triangles have **line symmetry**. There is another name for line symmetry. It is **reflection symmetry**.

Sample Reflections

Look at the figures below. In each one, the image on one side of the line is the original. The image on the other side of the line is the reflection. Notice the congruence of the two sides. See that the second image is a "flipped" version of the original.

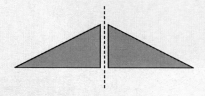 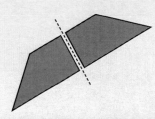 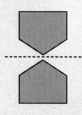

Reflecting an Image on a Grid

Look at the left triangle on the grid below. How do we find its reflection?

First, draw a line of reflection. In this case, the line of reflection is already drawn.

Next, choose a point. Find the distance between that point and the line of reflection. Point A is five units to the left of the line of reflection. That means that the reflected point will be five units to the *right* of the line. Plot point A' five units to the right of the line of reflection. Make sure you put it at the same height as point A.

Point B is three units to the left of the line. That means that the reflected point will be three units to the *right* of the line. Plot point B' three units to the right of the line of reflection.

Point C is one unit to the left of the line. That means that the reflected point will be one unit to the *right* of the line. Plot point C' one unit to the right of the line of reflection. Connect all the points.

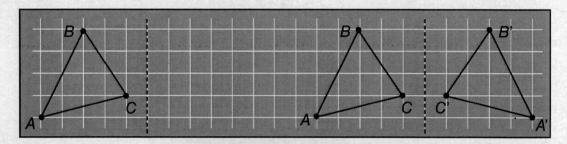

Reflections in Our Daily Lives

At the start of the lesson you thought about what words look like in a mirror. Each is a reflection of the original. Have you ever noticed the front of an ambulance? The word *ambulance* is a reflection. It is flipped. This way, a driver in front of an oncoming ambulance can read the word in his or her rearview mirror.

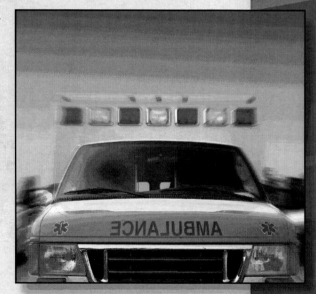

You Try It

Draw a reflection of the figure below.

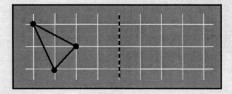

Reflections

Have you ever looked at words in a mirror? Are those words easy to read or do they seem backwards?

Basic Facts

A **transformation** is an adjustment of a figure. The figure can be the same size and shape but in a different position, or the figure can be stretched or shrunk.

One type of transformation is the **reflection**. A figure is reflected if it is flipped. The line that the figure is flipped along is called the **line of reflection**. The original figure and the reflected figure are the same size and shape, which means that they are congruent. Corresponding points are the same distance from the line of reflection. For example, each point A is five units from the line of reflection, each point B is one unit from the line of reflection, and each point C is three units from the line of reflection.

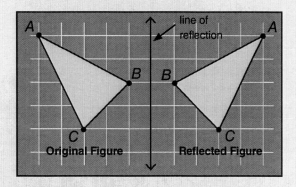

Look at the two triangles above. Do you see the symmetry in each picture? The triangles have **line symmetry**. Another name for line symmetry is **reflection symmetry**.

Sample Reflections

In each example below, the image on one side of the line is the original, while the image on the other side of the line is the reflection. Notice the congruence of the two sides. See that the second image is a "flipped" version of the original.

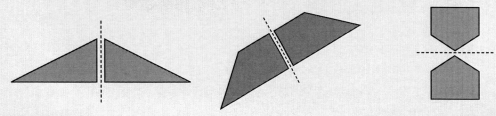

Reflecting an Image on a Grid

The left triangle on the grid below is to be reflected, but how do we do this?

First, draw a line of reflection. In this case, the line of reflection is already given.

Next, choose a point and find the distance between that point and the line of reflection. Point *A* is five units to the left of the line of reflection, so the reflected point will be five units to the *right* of the line. Plot point *A'* five units to the right of the line of reflection. Be sure it is on the same horizontal line.

Point *B* is three units to the left of the line, so the reflected point will be three units to the *right* of the line. Plot point *B'* three units to the right of the line of reflection.

Point *C* is one unit to the left of the line, so the reflected point will be one unit to the *right* of the line. Plot point *C'* one unit to the right of the line of reflection. Connect all the points.

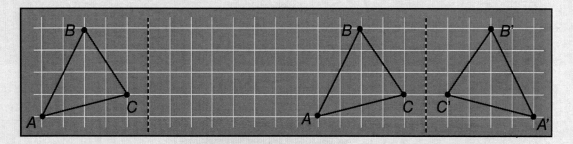

Reflections in Our Daily Lives

At the beginning of the lesson you thought about what words look like in a mirror. Each is a reflection of the original. Have you ever noticed the front of an ambulance? The word *ambulance* is a reflection. It is flipped. This way, a driver in front of an oncoming ambulance can read the word correctly in his or her rearview mirror.

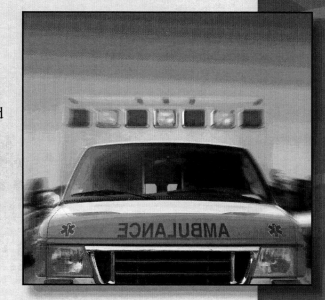

You Try It

Draw a reflection of the figure below.

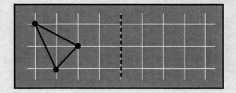

Reflections

Have you ever looked at words in a mirror? Are the words easy to read or do they appear backwards?

Basic Facts

A **transformation** is an adjustment of a figure. The figure can be the same size and shape but in a different position, or the figure can be stretched or shrunk.

One type of transformation is a **reflection**. This is a figure that is flipped. The line that the figure is flipped across is called the **line of reflection**. The original figure and the reflected figure are the same size and shape, which means that they are congruent. Corresponding points are the same distance from the line of reflection. For example, each point A in the figure below is five units from the line of reflection, each point B is one unit from the line of reflection, and each point C is three units from the line of reflection.

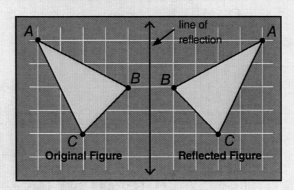

Examine the two triangles above for the **line symmetry** in each one. Another name for line symmetry is **reflection symmetry**.

Sample Reflections

In each example below, the image on one side of the line is the original, while the image on the other side of its line is the reflection. Notice the congruence of the two sides and that the second image is a "flipped" version of the original.

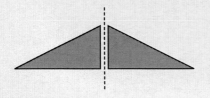 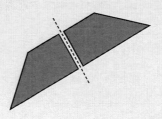 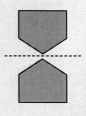

Reflecting an Image on a Grid

How would we find the reflection of the left triangle on the grid below?

First, draw a line of reflection, which, in this case, is already given.

Next, choose a point and find the distance between that point and the line of reflection. Point *A* is five units to the left of the line of reflection, so the reflected point will be five units to the *right* of the line. Plot point *A'* five units to the right of the line of reflection. Make sure you keep it at the same level horizontally.

Point *B* is three units to the left of the line, so the reflected point will be three units to the *right* of the line. Plot point *B'* three units to the right of the line of reflection.

Point *C* is one unit to the left of the line, so the reflected point will be one unit to the *right* of the line. Plot point *C'* one unit to the right of the line of reflection. Finally, connect all the points.

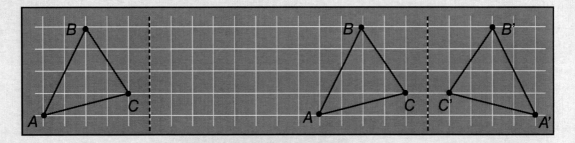

Reflections in Our Daily Lives

At the beginning of the lesson, we discussed what words look like when they are reflected in a mirror. Have you ever noticed that the word *ambulance* on the front of the emergency vehicle is flipped? This allows a driver in front of an oncoming ambulance to read the word correctly in his or her rearview mirror.

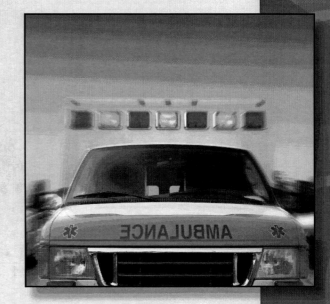

You Try It

Draw a reflection of the figure below.

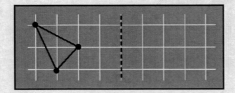

Rotations

Do you like rides? Do you like slow rides like the Ferris wheel? Or do you like rides that spin you around fast? Is there one that you like best?

Basic Facts

A transformation is when a figure is changed. The figure can be the same size and shape as the original. The position might change. Or the figure can be stretched or shrunk.

One type of transformation is the **rotation**. This can also be called a **turn**. When a figure is rotated, it is moved around a fixed point. This point is called the **center of rotation**. The figure and the fixed point lie in the same plane.

 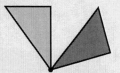

Before Rotation   Both Figures   After Rotation

Think about the Ferris wheel. Now think of the ride that spins fast. Both of them rotate you around a fixed point. You are the figure being turned around the center!

Rotational Symmetry

Some shapes can be rotated with no change in the way they look. Those shapes have **rotational symmetry**. Each figure below has rotational symmetry. Place your finger on the center of the first figure. Turn the paper upside down. Now the figure should look just like the original. You can try this with the second figure with 120° turns. You can try this with the third figure with 90° turns. You can try this with the last figure with 72° turns.

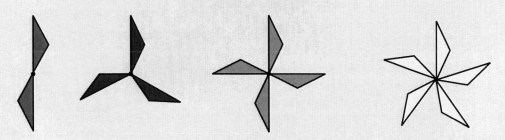

Rotations with Various Center Points

Key Details to Know

The center of rotation can be anywhere in the plane. It can even be on the figure. You can measure turns. You can use angle measurements. Phrases like "half-turn" may also be used. If you turn the number 6 with a half-turn, you get the number 9.

Rotations have direction. A turn may be clockwise ↻.

It may be counterclockwise ↺.

A turn can be described by the angle of rotation. It can also be described by the direction.

Looking at Some Examples

Each figure below shows a rotation. We start out with the dashed figure. The solid figure is the result after the rotation. Each turn is 120° clockwise. But each one uses a different center as the rotating point.

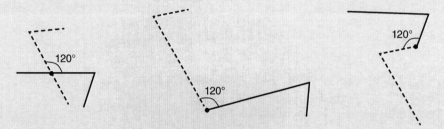

Rotations in Our Daily Lives

Look at a clock. Turn a doorknob. Turn on a fan. Rotations are a big part of your life. Patterns with flowers are used in many homes. Think of pillows or curtains. Often, the flowers have rotational symmetry. How do the designers know the size and numbers of petals to use? They use what they know about angles, circles, and rotational symmetry!

You Try It

Draw a triangle. Rotate it 90°. Now rotate it 180°.

Rotations

Do you like rides? Do you like slow rides like the Ferris wheel? Do you like rides that spin fast? Is there one that you like best?

Basic Facts

A transformation is when a figure is changed. The figure can be the same size and shape. The position might change. The figure can be stretched. It can be smaller.

One type of transformation is the **rotation**. This can also be called a **turn**. When a figure is rotated, it is moved around a fixed point. This point is called the **center of rotation**. The figure and the fixed point lie in the same plane.

 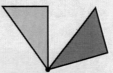

Before Rotation   Both Figures   After Rotation

Think about the Ferris wheel. Now think of the ride that spins fast. Both of them rotate the rider around a fixed point. The rider is the figure being turned around the center!

Rotational Symmetry

Sometimes a shape can be rotated with no change in the way it looks. That shape has rotational symmetry. Each figure below has **rotational symmetry**. Place your finger on the center of the first figure. Turn the paper upside down. The figure should look just like the original. Try this with the second figure with 120° turns. Try this with the third figure with 90° turns. Try it with the last figure with 72° turns.

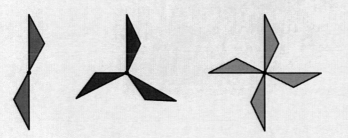 

Rotations with Various Center Points

Key Details to Know

The center of rotation can be anywhere in the plane. It can even be on the figure. You can measure turns. You can use angle measurements. Phrases like "half-turn" help describe the rotation. If you turn the number 6 with a half-turn, you get the number 9.

Rotations have direction. A turn may be clockwise ↻.

It may be counterclockwise ↺.

A turn can be described by the angle of rotation. It can also be described by the direction.

Looking at Some Examples

Each figure below shows a rotation. Start with the dashed line. The solid line shows the result after the rotation. Each turn is 120° clockwise. Each one uses a different center of rotation.

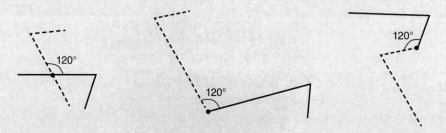

Rotations in Our Daily Lives

Look at a clock. Turn a doorknob. Turn on a fan. Rotations are a big part of your life. Many patterns have flowers as decoration. Think of pillows. Think of curtains. Often, the flowers have rotational symmetry. How does the artist choose the size of the flower? How does the artist choose the amount of petals? Artists use what they know about angles, circles, and rotational symmetry!

You Try It

Draw a triangle. Rotate it 90°. Now rotate it 180° from its original position.

Rotations

Would you like to ride a slow Ferris wheel? Or are you more in the mood for a fast, spinning ride that will send you upside down? Which type of ride do you like best?

Basic Facts

A transformation is an adjustment of a figure. The figure can be the same size and shape but in a different position, or the figure can be stretched or shrunk.

One type of transformation is the **rotation.** Sometimes this is also called a **turn.** When a figure is rotated, it is moved around a single, fixed point called the **center of rotation**. The figure and the fixed point lie in the same plane.

 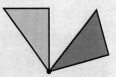

Before Rotation   Both Figures   After Rotation

Imagine being on the Ferris wheel or the fast, spinning ride. In either case, you are rotated around a single, fixed point. On either ride, you are a figure being rotated around the center.

Rotational Symmetry

When an entire shape can be rotated with no change in the way it looks, then that shape has **rotational symmetry**. Each figure below has rotational symmetry. Place your finger on the center of the first figure. Turn the paper upside down. The resulting figure should look just like the original. You can try this with the second figure with 120° turns. You can try this with the third figure with 90° turns. You can try this with the last figure with 72° turns.

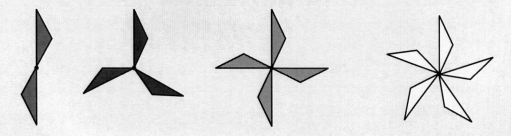

Rotations with Various Center Points

Key Details to Know

The center of rotation may be anywhere in the plane—even on the figure itself.

Turns can be measured with angle measurements. Phrases like "half-turn" may also be used. For example, if you turn the number 6 with a half-turn, you get the number 9.

Rotations have direction. A turn may be clockwise ↻ or counterclockwise ↺.

When a figure is rotated about a point, the turn may be described by the angle of rotation and by the direction.

Looking at Some Examples

Each figure below shows a rotation. The dashed figure is the figure we started with. The solid figure is the result after the rotation. Each turn is 120° clockwise, but each one uses a different center of rotation.

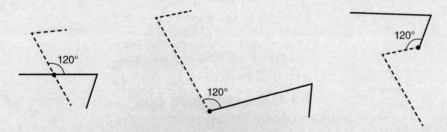

Rotations in Our Daily Lives

Look at a clock, turn a doorknob, or turn on a fan. Rotations are a part of your life every day. Think of bedspreads or curtains with flowers. Floral patterns are used in home decorating. Often these flowers have rotational symmetry. How do the designers know the size and numbers of petals to use? They use knowledge of angles, circles, and rotational symmetry!

You Try It

Draw a triangle. Rotate it 90° and 180° from its original position.

Rotations

When you go to an amusement park, do you enjoy riding the slow Ferris wheel? Or do you prefer fast, spinning rides that flip you upside down?

Basic Facts

A transformation is an adjustment of a figure. The figure can be the same size and shape as the original, but it can be in a different position, or the figure can be stretched or shrunk.

One type of transformation is the **rotation**, which can also be called a **turn**. When a figure is rotated, it is moved around a single, fixed point called the **center of rotation**.

 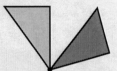

Before Rotation   Both Figures   After Rotation

Imagine being on the Ferris wheel or the faster, spinning ride. In both cases, you become the figure being rotated around a single, fixed point in the center.

Rotational Symmetry

When an entire shape can be rotated with no change in its appearance, then that shape has **rotational symmetry**. Each figure below has rotational symmetry. Place your finger on the center of the first figure and turn the paper upside down. The resulting figure should look just like the original. You can try this with the second figure with 120° turns, the third figure with 90° turns, and the last figure with 72° turns.

Rotations with Various Center Points

Key Details to Know

The center of rotation may be anywhere in the plane—even on the figure itself.

Turns can be measured with angle measurements. Phrases like "half-turn" may also be used. For example, if you turn the number 6 with a half-turn, you get the number 9.

Rotations have direction. A turn may be clockwise ↻ or counterclockwise ↺.

When a figure is rotated about a point, the turn may be described by the angle of rotation and by the direction.

Looking at Some Examples

Each figure shows a rotation. The dashed figure is the figure we started with, while the solid figure is the result after the rotation. Each turn is 120° clockwise, but each uses a different center of rotation.

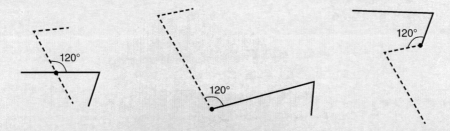

Rotations in Our Daily Lives

Look at a clock, turn a doorknob, or turn on a fan. Rotations are a part of your everyday life. You might see floral patterns on bedspreads or curtains, and often these patterns have rotational symmetry. How do the designers know the size and numbers of petals to use on the flowers? They use their knowledge of angles, circles, and rotational symmetry!

You Try It

Draw a triangle. Rotate it 90° and 180° from its original position.

#50717—Leveled Texts for Mathematics: Geometry © Shell Education

Translations

Would you like to go for a drive? When we drive, we use words such as *left*, *right*, *up*, and *down* to tell where we are going.

Basic Facts

A transformation is when there is a change to a figure. The figure can be the same size and shape. It might face a different way. Or it might be stretched or shrunk.

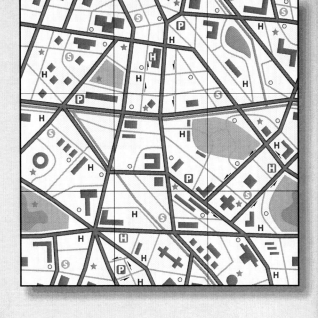

One type of transformation is the **translation**. It is also called a **slide**. A figure can be moved left or right. It can be moved up or down. In those cases, it is translated. The size or shape does not change. The first figure and the figure in the new place are congruent. Remember, congruent figures have the same size and shape.

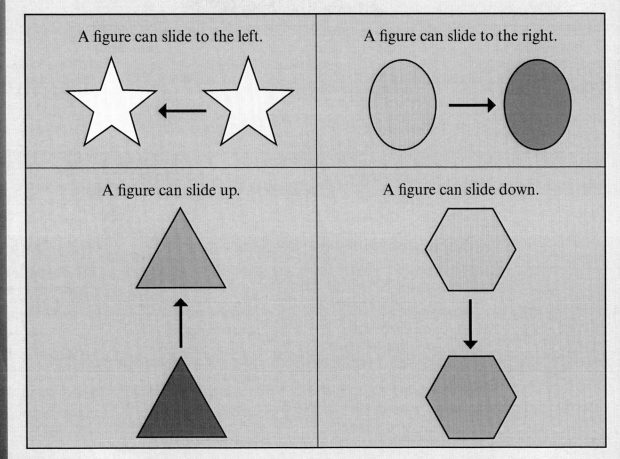

Graphing Translations

If a figure is drawn in a coordinate plane, translate key points to slide the entire figure.

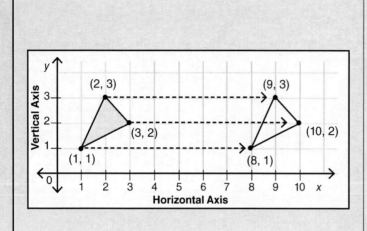

This graph shows a shaded triangle with vertices (1, 1), (3, 2), and (2, 3).

Look at (1, 1). It is moved 7 units to the right to its new location of (8, 1).

Look at (3, 2). It is moved 7 units to the right to its new location of (10, 2).

Look at (2, 3). It is moved 7 units to the right to its new location of (9, 3).

Each point was translated 7 units to the right. Each point was slid the same direction and by the same amount as every other point. We can say that the shaded triangle was slid 7 units to the right. We can connect the points of the new triangle. When we do, we have two congruent triangles.

Translations in Our Daily Lives

Do you like to ride in elevators? You move up and down. You can be seen as translation, or a slide. Now think about an escalator. You slide both vertically and horizontally.

You Try It

Translate the triangle below 6 units to the right. Draw the new triangle.

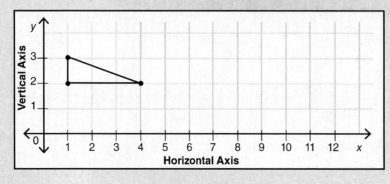

Translations

Describe how you might drive to get from start to finish. Use words such as *left*, *right*, *up*, and *down*.

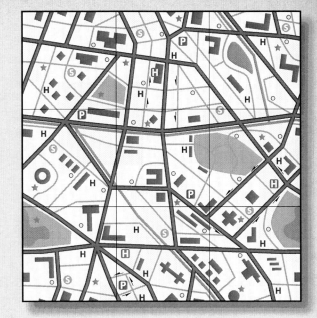

Basic Facts

A transformation is an adjustment of a figure. The figure can be the same size and shape. It might be in a different position. Or it might be stretched or shrunk.

One type of transformation is the **translation**. It can also be called a **slide**. A figure can be moved left, right, up, or down. In those cases, it is translated. The movement does not change the size or shape of the figure. The first figure and the figure in the new place are congruent. Remember, congruent figures have the same size and shape.

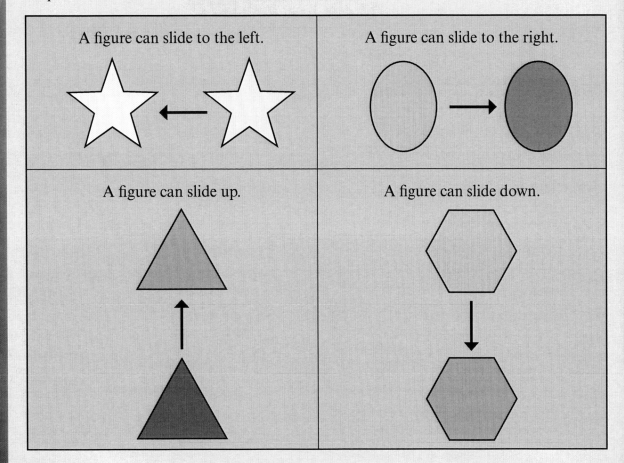

© Shell Education　　　　#50717—Leveled Texts for Mathematics: Geometry

Graphing Translations

If a figure is drawn in a coordinate plane, translate key points to slide the entire figure.

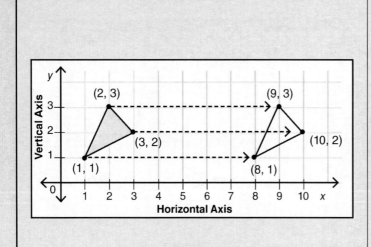

This graph shows a shaded triangle with vertices (1, 1), (3, 2), and (2, 3).

Look at (1, 1). It is moved 7 units to the right to its new location of (8, 1).

Look at (3, 2). It is moved 7 units to the right to its new location of (10, 2).

Look at (2, 3). It is moved 7 units to the right to its new location of (9, 3).

Each point was translated 7 units to the right. Each point was slid the same direction and by the same amount as every other point. We can say that the shaded triangle was slid 7 units to the right. We can connect the points of the new triangle. When we do, we have two congruent triangles.

Translations in Our Daily Lives

Do you like to ride in elevators? When you do, your movement can be seen as a translation, or a slide. You move up and down. Now think about an escalator. Your position slides both vertically and horizontally.

You Try It

Translate the triangle below 6 units to the right. Draw the resulting triangle.

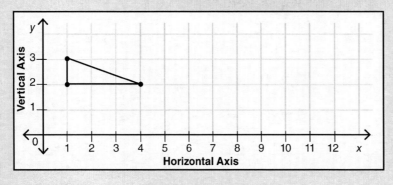

Translations

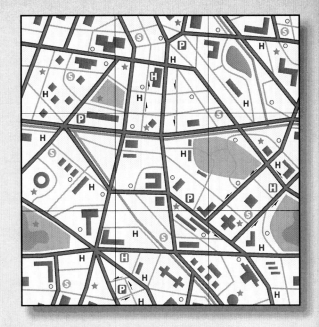

Describe how you might drive to get from start to finish, using words such as *left*, *right*, *up*, and *down*.

Basic Facts

A transformation is an adjustment of a figure. The figure can be the same size and shape but in a different position, or the figure can be stretched or shrunk.

One type of transformation is the **translation**. Sometimes this is also called a **slide**. When a figure is moved left, right, up, or down, it is translated. The movement does nothing to the size or shape of the figure being moved. The figure in the original position and the figure in the moved position are congruent. Remember, congruent figures have the same size and shape.

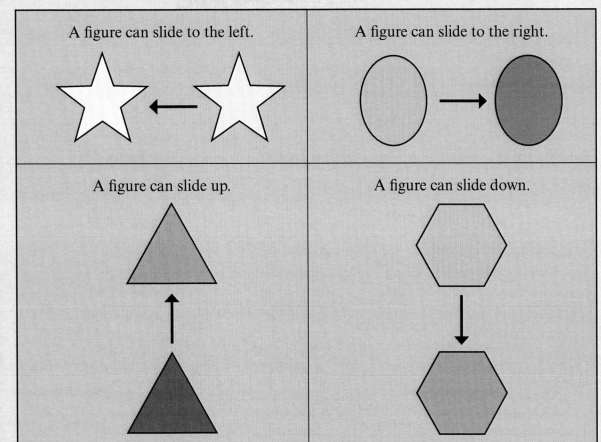

Graphing Translations

If a figure is drawn in a coordinate plane, translate key points to slide the entire figure.

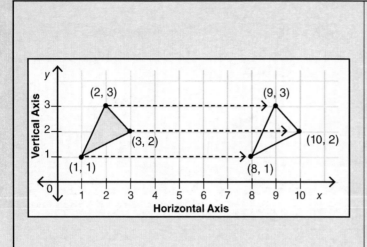

This graph shows a shaded triangle with vertices (1, 1), (3, 2), and (2, 3).

Look at (1, 1). It is moved 7 units to the right to its new location of (8, 1).

Look at (3, 2). It is moved 7 units to the right to its new location of (10, 2).

Look at (2, 3). It is moved 7 units to the right to its new location of (9, 3).

Each point was translated seven units to the right. Each point was slid the same direction and by the same amount as every other point, so we can say that the shaded triangle was slid seven units to the right. When the points of the new triangle are connected, the result is two congruent triangles.

Translations in Our Daily Lives

If you ride in an elevator, your movement can be seen as a translation, or a slide. You move up and down. On an escalator, your position slides both vertically and horizontally.

You Try It

Translate the triangle below 6 units to the right. Draw the resulting triangle.

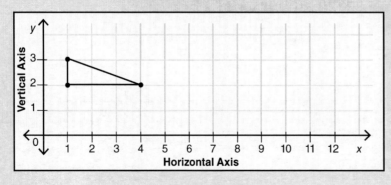

Translations

Describe how you might drive to get from your starting point to your destination, using words such as *left*, *right*, *up*, and *down*.

Basic Facts

A transformation is an adjustment of a figure. The figure can be the identical size and shape as the original. However, it can be in a different position, or the figure can be stretched or shrunk.

One particular type of transformation is the **translation**, which can also be called a **slide**. When a figure is moved left, right, up, or down, it is translated. The movement does nothing to alter the size or shape of the figure being moved, and the figure in the original position and the figure in the new position remain congruent. Remember, congruent figures have the same size and shape as one another.

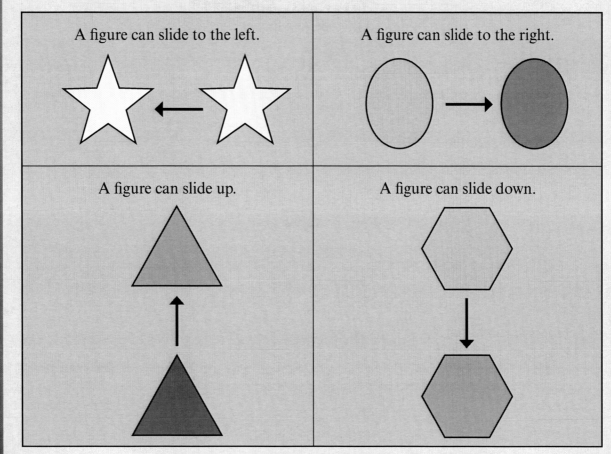

Graphing Translations

If a figure is drawn in a coordinate plane, translate key points to slide the entire figure.

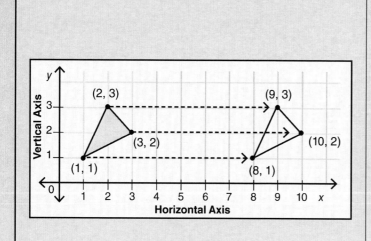

This graph shows a shaded triangle with vertices (1, 1), (3, 2), and (2, 3).

Look at (1, 1). It is moved 7 units to the right to its new location of (8, 1).

Look at (3, 2). It is moved 7 units to the right to its new location of (10, 2).

Look at (2, 3). It is moved 7 units to the right to its new location of (9, 3).

Each point was translated 7 units to the right. Each point was slid the same direction and exactly by the same amount as every other point, so we can conclude that the shaded triangle was slid 7 units to the right. When the points of the new triangle are connected, the first triangle and the one just drawn are congruent.

Translations in Our Daily Lives

When you ride an elevator, you move up and down, so your movement can be described as a translation, or a slide. On an escalator, your position slides both vertically and horizontally.

You Try It

Translate the triangle below 6 units to the right and draw the resulting triangle.

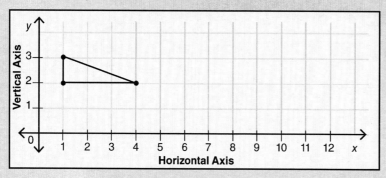

References Cited

August, D. and T. Shanahan (Eds). 2006. Developing literacy in second-language learners: Report of the National Literacy Panel on language-minority children and youth. Mahwah, NJ: Lawrence Erlbaum Associates, Inc.

Common Core State Standards Initiative. 2010. *The standards: Language arts.* (Accessed October 2010.) http://www.corestandards.org/the-standards/languagearts.

Marzano, R., D. Pickering, and J. Pollock. 2001. *Classroom instruction that works.* Alexandria, VA: Association for Supervision and Curriculum Development.

Tomlinson, C.A. 2000. *Leadership for Differentiating Schools and Classrooms.* Alexandria, VA: Association for Supervision and Curriculum Development.

Vygotsky, L.S. 1978. *Mind and society: The development of higher mental processes.* Cambridge, MA: Harvard University Press.

Contents of Teacher Resource CD

NCTM Mathematics Standards

The National Council of Teachers of Mathematics (NCTM) standards are listed in the chart on page 20, as well as on the Teacher Resource CD: *nctm.pdf*. Also included are TESOL Standards: *TESOL.pdf*.

Text Files

The text files include the text for all four levels of each reading passage. For example, the Angles All Around text (pages 21–28) is the *angles_all_around.doc* file.

PDF Files

The full-color PDFs provided are each eight pages long and contain all four levels of a reading passage. For example, the Angles All Around PDF (pages 21–28) is the *angles_all_around.pdf* file.

Text Title	Text File	PDF
Angles All Around	angles_all_around.doc	angles_all_around.pdf
Understanding Triangles	triangles.doc	triangles.pdf
To Cross or Not to Cross	to_cross.doc	to_cross.pdf
Quadrilaterals	quadrilaterals.doc	quadrilaterals.pdf
Classifying 2-D Shapes	classifying.doc	classifying.pdf
Irregular Shapes	irregular_shapes.doc	irregular_shapes.pdf
Congruent and Similar Figures	congruent_and_similar.doc	congruent_and_similar.pdf
Understanding 3-D Shapes	3-D_shapes.doc	3-D_shapes.pdf
Understanding Prisms	prisms.doc	prisms.pdf
The Coordinate Plane	coordinate_plane.doc	coordinate_plane.pdf
Circles	circles.doc	circles.pdf
Symmetry	symmetry.doc	symmetry.pdf
Reflections	reflections.doc	reflections.pdf
Rotations	rotations.doc	rotations.pdf
Translations	translations.doc	translations.pdf

JPEG Files

Key mathematical images found in the book are also provided on the Teacher Resource CD.

Word Documents of Texts
- Change leveling further for individual students.
- Separate text and images for students who need additional help decoding the text.
- Resize the text for visually impaired students.

Full-Color PDFs of Texts
- Create overhead transparencies or color copies to display on a document projector.
- Project texts on an interactive whiteboard or other screen for whole-class review.
- Read texts online.
- Email texts to parents or students at home.

JPEGs of Mathematical Images
- Display as visual support for use with whole class or small-group instruction.

Notes

Notes